BIG IDEAS
MATH
Blue

A Common Core Curriculum

Record and Practice Journal

- Fair Game Review Worksheets

- Activity Recording Journal

- Practice Worksheets

- Glossary

- Activity Manipulatives

BIG IDEAS LEARNING

Erie, Pennsylvania

About the Record and Practice Journal

Fair Game Review

The Fair Game Review corresponds to the Pupil Edition Chapter Opener. Here you have the opportunity to practice prior skills necessary to move forward.

Activity Recording Journal

The Activity pages correspond to the Activity in the Pupil Edition. Here you have room to show your work and record your answers.

Practice Worksheets

Each section of the Pupil Edition has an additional Practice page with room for you to show your work and record your answers.

Glossary

This student-friendly glossary is designed to be a reference for key vocabulary, properties, and mathematical terms. Several of the entries include a short example to aid your understanding of important concepts

Activity Manipulatives

Manipulatives needed for the activities are included in the back of the Record and Practice Journal.

Contents

Contents

Contents

Contents

Contents

Contents

Chapter 1 — Fair Game Review

Simplify the expression.

1. $18x - 6x + 2x$

2. $4b - 7 - 15b + 3$

3. $15(6 - g)$

4. $-24 + 2(y - 9)$

5. $9m + 4(12 - m)$

6. $16(a - 2) + 3(10 - a)$

7. You are selling lemonade for $1.50, a bag of kettle corn for $3, and a hot dog for $2.50 at a fair. Write and simplify an expression for the amount of money you receive when p people buy one of each item.

Chapter 1 **Fair Game Review** (continued)

Add or subtract.

8. $-1 + (-3)$

9. $0 + (-12)$

10. $-5 + (-3)$

11. $-4 + (-4)$

12. $5 - (-2)$

13. $-5 - 2$

14. $0 - (-6)$

15. $-9 - 3$

16. In a city, the record monthly high temperature for July is $88°F$. The record monthly low temperature is $30°F$. What is the range of temperatures for July?

Name_____ Date_____

1.1 Solving Simple Equations
For use with Activity 1.1

Essential Question How can you use inductive reasoning to discover rules in mathematics? How can you test a rule?

1 ACTIVITY: Sum of the Angles of a Triangle

Work with a partner. Use a protractor to measure the angles of each triangle. Complete the table to organize your results.

a.
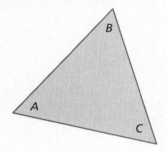

b.
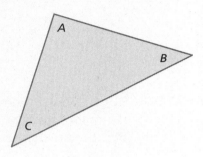

c.
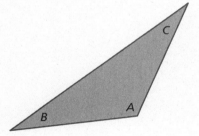

d.
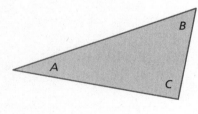

Triangle	Angle *A* (degrees)	Angle *B* (degrees)	Angle *C* (degrees)	*A* + *B* + *C*
a.				
b.				
c.				
d.				

Name _____ Date _____

Solving Simple Equations (continued)

2 **ACTIVITY:** Writing a Rule

Work with a partner. Use inductive reasoning to write and test a rule.

    a. STRUCTURE Use the completed table in Activity 1 to write a rule about the sum of the angle measures of a triangle.

    b. TEST YOUR RULE Draw four triangles that are different from those in Activity 1. Measure the angles of each triangle. Organize your results in a table. Find the sum of the angle measures of each triangle.

4 **Big Ideas Math Blue**
Record and Practice Journal

Copyright © Big Ideas Learning, LLC

1.1 Solving Simple Equations (continued)

3 **ACTIVITY:** Applying Your Rule

Work with a partner. Use the rule you wrote in Activity 2 to write an equation for each triangle. Then solve the equation to find the value of *x*. Use a protractor to check the reasonableness of your answer.

a.
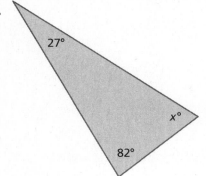

27°

x°

82°

b.
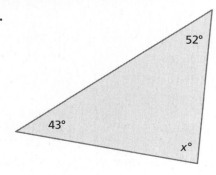

52°

43°

x°

c.
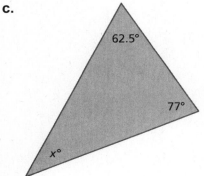

62.5°

77°

x°

d.
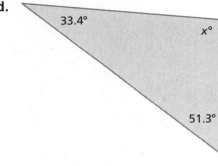

33.4°

x°

51.3°

What Is Your Answer?

4. **IN YOUR OWN WORDS** How can you use inductive reasoning to discover rules in mathematics? How can you test a rule? How can you use a rule to solve problems in mathematics?

Big Ideas Math Blue
Record and Practice Journal

5

1.1 Practice
For use after Lesson 1.1

Solve the equation. Check your solution.

1. $x + 5 = 16$

2. $11 = w - 12$

3. $\dfrac{3}{4} + z = \dfrac{5}{6}$

4. $3y = 18$

5. $\dfrac{k}{7} = 10$

6. $\dfrac{4}{5}n = \dfrac{9}{10}$

7. $x - 12 \div 6 = 9$

8. $h + |-8| = 15$

9. $1.3(2) + p = 7.9$

10. A coupon subtracts \$5.16 from the price p of a shirt. You pay \$15.48 for the shirt after using the coupon. Write and solve an equation to find the original price of the shirt.

1.2 Solving Multi-Step Equations
For use with Activity 1.2

Essential Question How can you solve a multi-step equation? How can you check the reasonableness of your solution?

---

1 ACTIVITY: Solving for the Angles of a Triangle

Work with a partner. Write an equation for each triangle. Solve the equation to find the value of the variable. Then find the angle measures of each triangle. Use a protractor to check the reasonableness of your answer.

a.
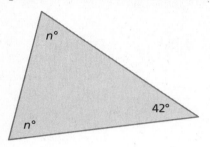

b.
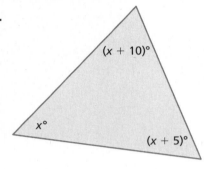

c.
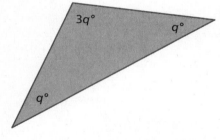

d.
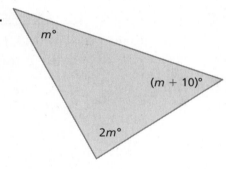

e.
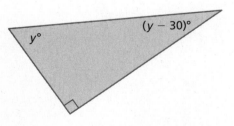

f.
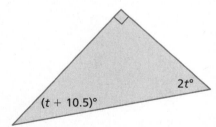

1.2 **Solving Multi-Step Equations** (continued)

2 **ACTIVITY:** Problem Solving Strategy

Work with a partner.

The six triangles form a rectangle.

Find the angle measures of each triangle. Use a protractor to check the reasonableness of your answers.

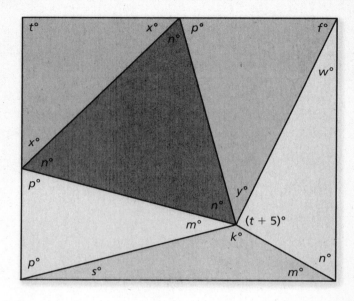

3 **ACTIVITY:** Puzzle

Work with a partner. A survey asked 200 people to name their favorite weekday. The results are shown in the circle graph.

a. How many degrees are in each part of the circle graph?

b. What percent of the people chose each day?

c. How many people chose each day?

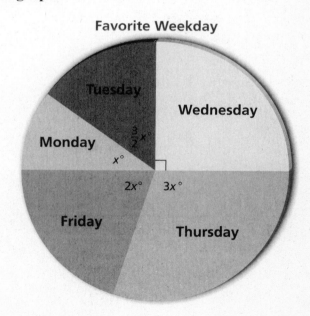

Favorite Weekday

1.2 Solving Multi-Step Equations (continued)

d. Organize your results in a table.

What Is Your Answer?

4. IN YOUR OWN WORDS How can you solve a multi-step equation?
How can you check the reasonableness of your solution?

1.2 Practice
For use after Lesson 1.2

Solve the equation. Check your solution.

1. $3x - 11 = 22$

2. $24 - 10b = 9$

3. $2.4z + 1.2z - 6.5 = 0.7$

4. $\dfrac{3}{4}w - \dfrac{1}{2}w - 4 = 12$

5. $2(a + 7) - 7 = 9$

6. $20 + 8(q - 11) = -12$

7. Find the width of the rectangular prism when the surface area is 208 square centimeters.

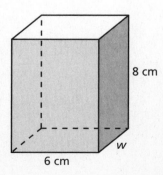

8 cm

6 cm

w

Name_____ Date_____

1.3 Solving Equations with Variables on Both Sides
For use with Activity 1.3

Essential Question How can you solve an equation that has variables on both sides?

1 ACTIVITY: Perimeter and Area

Work with a partner.

- Each figure has the unusual property that the value of its perimeter (in feet) is equal to the value of its area (in square feet). Write an equation for each figure.

- Solve each equation for x.

- Use the value of x to find the perimeter and the area of each figure.

- Describe how you can check your solution.

a.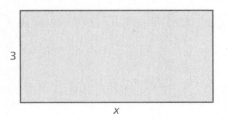

3

x

b.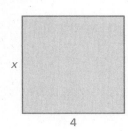

x

4

c.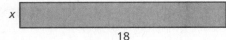

x

18

d.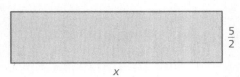

$\frac{5}{2}$

x

1.3 **Solving Equations with Variables on Both Sides** (continued)

e.

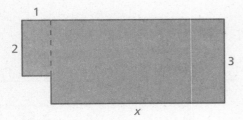

f.

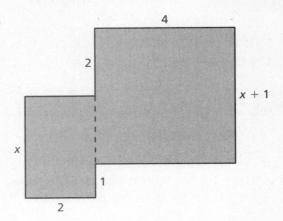

g.

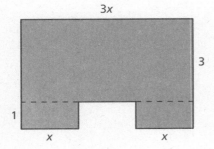

2 **ACTIVITY:** Surface Area and Volume

Work with a partner.

*   Each solid on the next page has the unusual property that the value of its surface area (in square inches) is equal to the value of its volume (in cubic inches). Write an equation for each solid.

*   Solve each equation for x.

*   Use the value of x to find the surface area and the volume of each solid.

*   Describe how you can check your solution.

1.3 **Solving Equations with Variables on Both Sides** (continued)

a.
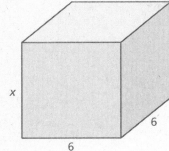

b.
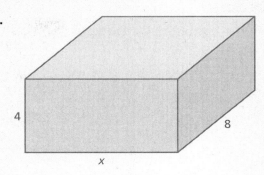

3 **ACTIVITY:** Puzzle

Work with a partner. The perimeter of the larger triangle is 150% of the perimeter of the smaller triangle. Find the dimensions of each triangle.

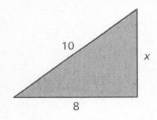

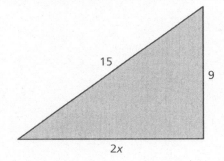

What Is Your Answer?

4. **IN YOUR OWN WORDS** How can you solve an equation that has variables on both sides? How do you move a variable term from one side of the equation to the other?

5. Write an equation that has variables on both sides. Solve the equation.

1.3 Practice
For use after Lesson 1.3

Solve the equation. Check your solution.

1. $x + 16 = 9x$

2. $4y - 70 = 12y + 2$

3. $5(p + 6) = 8p$

4. $3(g - 7) = 2(10 + g)$

5. $1.8 + 7n = 9.5 - 4n$

6. $\dfrac{3}{7}w - 11 = -\dfrac{4}{7}w$

7. One movie club charges a \$100 membership fee and \$10 for each movie. Another club charges no membership fee but movies cost \$15 each. Write and solve an equation to find the number of movies you need to buy for the cost of each movie club to be the same.

Name_____ Date_____

1.4 Rewriting Equations and Formulas
For use with Activity 1.4

Essential Question How can you use a formula for one measurement to write a formula for a different measurement?

---

1 ACTIVITY: Using Perimeter and Area Formulas

Work with a partner.

a. • Write a formula for the perimeter P of a rectangle.

   • Solve the formula for w.

   • Use the new formula to find the width of the rectangle.

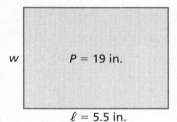

w $P = 19$ in.

$\ell = 5.5$ in.

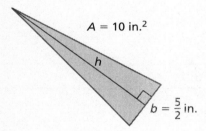

$A = 10$ in.^{2}

h

$b = \frac{5}{2}$ in.

b. • Write a formula for the area A of a triangle.

   • Solve the formula for h.

   • Use the new formula to find the height of the triangle.

c. • Write a formula for the circumference C of a circle.

   • Solve the formula for r.

   • Use the new formula to find the radius of the circle.

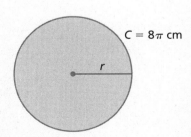

$C = 8\pi$ cm

r

1.4 **Rewriting Equations and Formulas** (continued)

- Write a formula for the area *A*.
- Solve the formula for *h*.
- Use the new formula to find the height.

d.

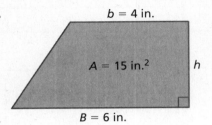

b = 4 in.

A = 15 in.²   h

B = 6 in.

e.

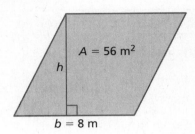

A = 56 m²

h

b = 8 m

---

2 **ACTIVITY:** Using Volume and Surface Area Formulas

Work with a partner.

a.
- Write a formula for the volume *V* of a prism.
- Solve the formula for *h*.
- Use the new formula to find the height of the prism.

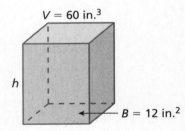

V = 60 in.³

h

B = 12 in.²

1.4 **Rewriting Equations and Formulas** (continued)

b. • Write a formula for the volume V of a pyramid.

• Solve the formula for B.

• Use the new formula to find the area of the base of the pyramid.

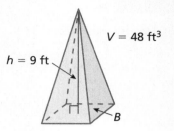

$V = 48$ ft^3

$h = 9$ ft

B

c. • Write a formula for the lateral surface area S of a cylinder.

• Solve the formula for h.

• Use the new formula to find the height of the cylinder.

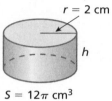

$r = 2$ cm

h

$S = 12\pi$ cm^3

d. • Write a formula for the surface area S of a rectangular prism.

• Solve the formula for ℓ.

• Use the new formula to find the length of the rectangular prism.

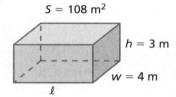

$S = 108$ m^2

$h = 3$ m

$w = 4$ m

ℓ

What Is Your Answer?

3. IN YOUR OWN WORDS How can you use a formula for one measurement to write a formula for a different measurement? Give an example that is different from the examples on these three pages.

1.4 Practice
For use after Lesson 1.4

Solve the equation for _y_.

1. $2x + y = -9$

2. $4x - 10y = 12$

3. $13 = \dfrac{1}{6}y + 2x$

Solve the formula for the bold variable.

4. $V = \ell w \mathbf{h}$

5. $f = \dfrac{1}{2}(\mathbf{r} + 6.5)$

6. $S = 2\pi r^2 + 2\pi r\mathbf{h}$

7. The formula for the area of a triangle is $A = \dfrac{1}{2}bh$.

    a. Solve the formula for _h_.

    b. Use the new formula to find the value of _h_.

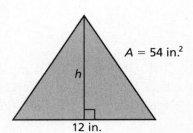

$A = 54 \text{ in.}^2$

12 in.

Name_____ Date_____

Chapter 2 Fair Game Review

Reflect the point in (a) the *x*-axis and (b) the *y*-axis.

1. $(1, 1)$

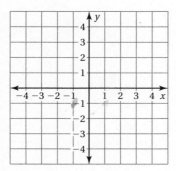

2. $(-2, -4)$

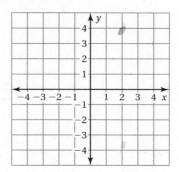

3. $(-3, 3)$

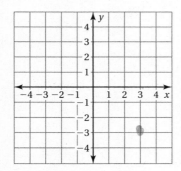

4. $(4, -3)$

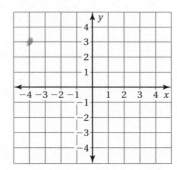

5. $(-1, 2)$

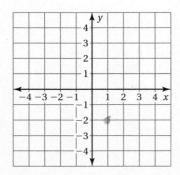

6. $(3, 2)$

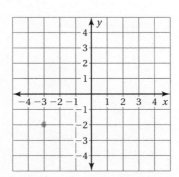

Chapter 2

Fair Game Review (continued)

Draw the polygon with the given vertices in a coordinate plane.

7. $A(2, 2)$, $B(2, 7)$, $C(6, 7)$, $D(6, 2)$

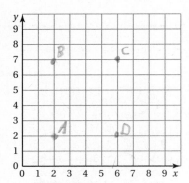

8. $E(3, 8)$, $F(3, 1)$, $G(6, 1)$, $H(6, 8)$

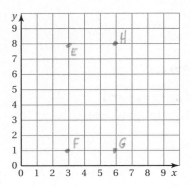

9. $I(7, 6)$, $J(5, 2)$, $K(2, 4)$

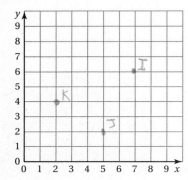

10. $L(1, 5)$, $M(1, 2)$, $N(8, 2)$

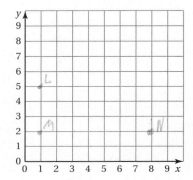

11. $O(3, 7)$, $P(6, 7)$, $Q(9, 3)$, $R(1, 3)$

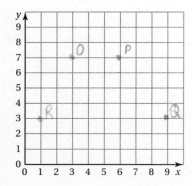

12. $S(9, 9)$, $T(7, 1)$, $U(2, 4)$, $V(4, 7)$

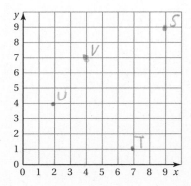

Name_____ Date_____

2.1 Congruent Figures
For use with Activity 2.1

Essential Question How can you identify congruent triangles?

Two figures are congruent when they have the same size and the same shape.

---

1 **ACTIVITY:** Identifying Congruent Triangles

Work with a partner.

- **Which of the geoboard triangles below are congruent to the geoboard triangle at the right?**

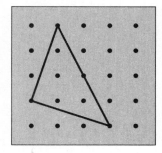

a.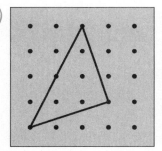

b.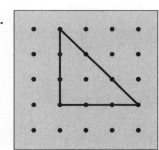

c.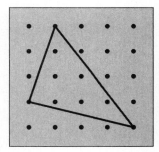

d.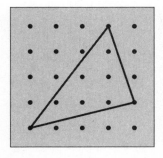

e.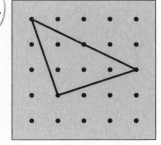

f.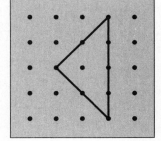

Name **Eric**   Date _____

2.1 **Congruent Figures** (continued)

- Form each triangle on a geoboard.

- Measure each side with a ruler. Record your results in the table.

	Side 1	Side 2	Side 3
Given Triangle	3		
a.	3 cm	2.3 cm	2.4 cm
b.	2.2 cm	2.1 cm	3 cm
c.	2.9 cm	3.2 cm	2 cm
d.	2.7 cm	2 cm	3.4
e.			
f.			

- Write a conclusion about the side lengths of triangles that are congruent.

   All   the sides   are   the
   Same   Size.

2.1 **Congruent Figures** (continued)

2 **ACTIVITY:** Forming Congruent Triangles

Work with a partner.

a. Form the given triangle in Activity 1 on your geoboard. Record the triangle on geoboard dot paper.

b. Move each vertex of the triangle one peg to the right. Is the new triangle congruent to the original triangle? How can you tell?

c. On a 5-by-5 geoboard, make as many different triangles as possible, each of which is congruent to the given triangle in Activity 1. Record each triangle on geoboard dot paper.

What Is Your Answer?

3. **IN YOUR OWN WORDS** How can you identify congruent triangles? Use the conclusion you wrote in Activity 1 as part of your answer.

The triangles are the same size and shape.

4. Can you form a triangle on your geoboard whose side lengths are 3, 4, and 5 units? If so, draw such a triangle on geoboard dot paper.

Name _____ Date _____

The figures are congruent. Name the corresponding angles and the corresponding sides.

1.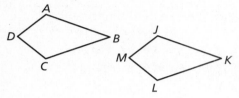

2.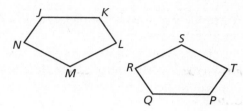

Tell whether the two figures are congruent. Explain your reasoning.

3.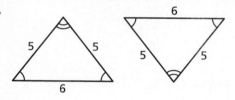

4.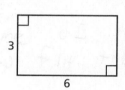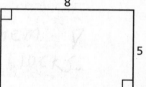

5. The tops of the desks are identical.

   a. What is the length of side *NP*?

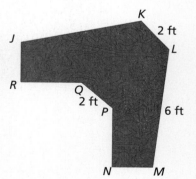

   b. Side *AB* is congruent to side *CD*. What is the length of side *AB*?

2.2 Translations
For use with Activity 2.2

Essential Question How can you arrange tiles to make a tessellation?

1 ACTIVITY: Describing Tessellations

Work with a partner. Can you make the tessellation by translating single tiles that are all of the same shape and design? If so, show how.

   a. Sample:

   Tile Pattern   **Single Tiles**

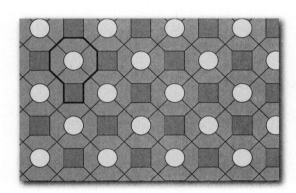
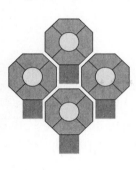

   b.   c.

2.2 **Translations** (continued)

2 **ACTIVITY:** Tessellations and Basic Shapes

Work with a partner.

  a. Which pattern blocks can you use to make a tessellation? For each one that works, draw the tessellation.

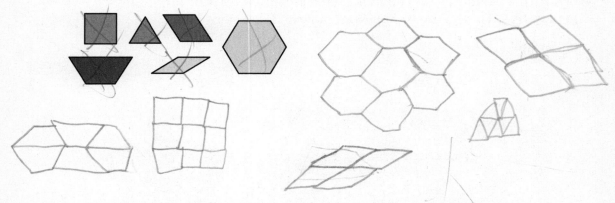

  b. Can you make the tessellation by translating? Or do you have to rotate or flip the pattern blocks? For some of them you have to flip the blocks.

3 **ACTIVITY:** Designing Tessellations

Work with a partner. Design your own tessellation. Use one of the basic shapes from Activity 2.

Sample:

Step 1: Start with a square.   **Step 2:** Cut a design out of one side.   **Step 3:** Tape it to the other side to make your pattern.

Step 4: Translate the pattern to make your tessellation.

Step 5: Color the tessellation.

2.2 **Translations** (continued)

4 **ACTIVITY:** Translating in the Coordinate Plane

Work with a partner.

   a. Draw a rectangle in a coordinate plane. Find the dimensions of the rectangle.

   b. Move each vertex 3 units right and 4 units up. Draw the new figure. List the vertices.

   c. Compare the dimensions and the angle measures of the new figure to those of the original rectangle.

   d. Are the opposite sides of the new figure parallel? Explain.

   e. Can you conclude that the two figures are congruent? Explain.

   f. Compare your results with those of other students in your class. Do you think the results are true for any type of figure?

What Is Your Answer?

   5. IN YOUR OWN WORDS How can you arrange tiles to make a tessellation? Give an example.

   6. PRECISION Explain why any parallelogram can be translated to make a tessellation.

Name _____ Date _____

Tell whether the shaded figure is a translation of the nonshaded figure.

1.

No

2.

Yes

3.

NO

4. Translate the figure 4 units left and 1 unit down. What are the coordinates of the image?

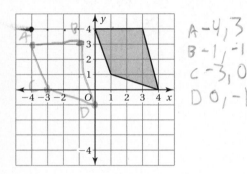

A −4, 3
B −1, −1
C −3, 0
D 0, −1

5. Translate the triangle 5 units right and 4 units up. What are the coordinates of the image?

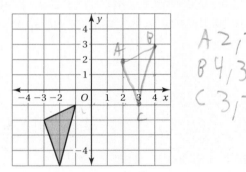

A 2, 2
B 4, 3
C 3, −1

6. Describe the translation from the shaded figure to the nonshaded figure.

translate the triangle 5 units down, and 3 units to the left.

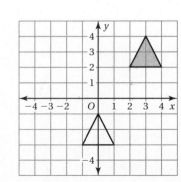

Name_____ Date _____

Essential Question How can you use reflections to classify a frieze pattern?

Frieze

A *frieze* is a horizontal band that runs at the top of a building. A frieze is often decorated with a design that repeats.

- All frieze patterns are translations of themselves.
- Some frieze patterns are reflections of themselves.

---

1 ACTIVITY: Frieze Patterns and Reflections

Work with a partner. Consider the frieze pattern shown. *

a. Is the frieze pattern a reflection of itself when folded horizontally? Explain.

Yes

b. Is the frieze pattern a reflection of itself when folded vertically? Explain.

yes

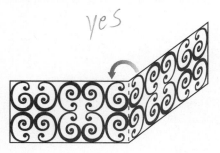

*Cut-outs are available in the back of the Record and Practice Journal.

Name _____ Date _____

2.3 **Reflections** (continued)

2 **ACTIVITY:** Frieze Patterns and Reflections

Work with a partner. Is the frieze pattern a reflection of itself when folded
horizontally, *vertically*, or *neither*?

a.

b.

Horizontal,

3 **ACTIVITY:** Reflecting in the Coordinate Plane

Work with a partner.

  a. Draw a rectangle in Quadrant I of a coordinate
     plane. Find the dimensions of the rectangle.

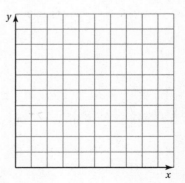

  b. Copy the axes and the rectangle onto a piece of
     transparent paper.

     Flip the transparent paper once so that the rectangle
     is in Quadrant IV. Then align the origin and the axes
     with the coordinate plane.

     Draw the new figure in the coordinate plane.
     List the vertices.

30 **Big Ideas Math Blue**
    Record and Practice Journal

2.3   **Reflections** (continued)

c. Compare the dimensions and the angle measures of the new figure to those of the original rectangle.

d. Are the opposite sides of the new figure still parallel? Explain.

e. Can you conclude that the two figures are congruent? Explain.

f. Flip the transparent paper so that the original rectangle is in Quadrant II. Draw the new figure in the coordinate plane. List the vertices. Then repeat parts (c)−(e).

g. Compare your results with those of other students in your class. Do you think the results are true for any type of figure?

What Is Your Answer?

4. IN YOUR OWN WORDS How can you use reflections to classify a frieze pattern?

Name _____ Date _____

Tell whether the shaded figure is a reflection of the nonshaded figure.

1.  NO

2. 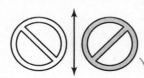 Yes

3. 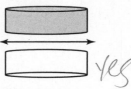 Yes

Draw the figure and its reflection in the *x*-axis. Identify the coordinates of the image.

4. $A(1, 2)$, $B(3, 2)$, $C(1, 4)$

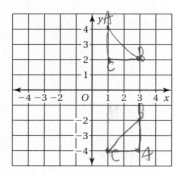

5. $W(3, 1)$, $X(3, 4)$, $Y(1, 4)$, $Z(1, 1)$

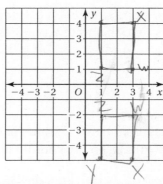

Draw the figure and its reflection in the *y*-axis. Identify the coordinates of the image.

6. $J(3, 4)$, $K(3, 0)$, $L(2, 4)$

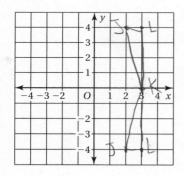

7. $M(2, 2)$, $N(2, 3)$, $P(3, 3)$, $Q(4, 1)$

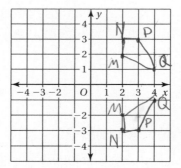

8. In a pinball game, when you perfectly reflect the ball off of the wall, will the ball hit the bonus target? NO

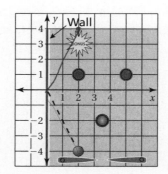

2.4 Rotations
For use with Activity 2.4

Essential Question What are the three basic ways to move an object in a plane?

1 ACTIVITY: Three Basic Ways to Move Things

There are three basic ways to move objects on a flat surface.

<u>slide</u> the object.   <u>reflected</u> the object.   <u>rotation</u> the object.

Work with a partner.

a. What type of triangle is the shaded triangle? Is it congruent to the other triangles? Explain.

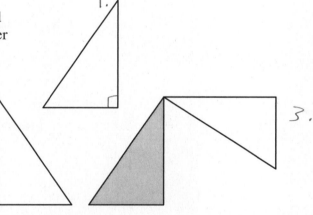

right triangle, and yes because the other triangles are just rotated, flipped, or reversed - all right triangles

b. Decide how you can move the shaded triangle to obtain each of the other triangles.
1. slide it
2. reflect it
3. rotate it

c. Is each move a *translation*, a *reflection*, or a *rotation*?

Name _____ Date _____

2.4 **Rotations** (continued)

2 **ACTIVITY:** Rotating in the Coordinate Plane

Work with a partner.

a. Draw a rectangle in Quadrant II of a coordinate plane. Find the dimensions of the rectangle.

b. Copy the axes and the rectangle onto a piece of transparent paper.

Align the origin and the vertices of the rectangle on the transparent paper with the coordinate plane. Turn the transparent paper so that the rectangle is in Quadrant I and the axes align.

Draw the new figure in the coordinate plane. List the vertices.

c. Compare the dimensions and the angle measures of the new figure to those of the original rectangle.

d. Are the opposite sides of the new figure still parallel? Explain.

e. Can you conclude that the two figures are congruent? Explain.

f. Turn the transparent paper so that the original rectangle is in Quadrant IV. Draw the new figure in the coordinate plane. List the vertices. Then repeat parts (c)–(e).

2.4 **Rotations** (continued)

g. Compare your results with those of other students in your class. Do you think the results are true for any type of figure?

What Is Your Answer?

3. IN YOUR OWN WORDS What are the three basic ways to move an object in a plane? Draw an example of each.

4. PRECISION Use the results of Activity 2(b).

  a. Draw four angles using the conditions below.

  - The origin is the vertex of each angle.
  - One side of each angle passes through a vertex of the original rectangle.
  - The other side of each angle passes through the corresponding vertex of the rotated rectangle.

  b. Measure each angle in part (a). For each angle, measure the distances between the origin and the vertices of the rectangles. What do you notice?

  c. How can the results of part (b) help you rotate a figure?

5. PRECISION Repeat the procedure in Question 4 using the results of Activity 2(f).

2.4 **Practice**
For use after Lesson 2.4

Tell whether the shaded figure is a rotation of the nonshaded figure about the origin. If so, give the angle and the direction of rotation.

1.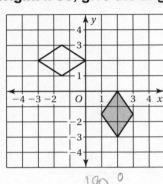

180°
clock wise

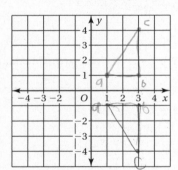

clockwise

2.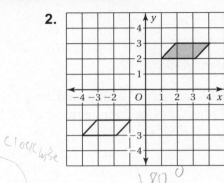

180°
counter
clock wise

The vertices of a triangle are $A(1, 1)$, $B(3, 1)$, **and** $C(3, 4)$. **Rotate the triangle as described. Find the coordinates of the image.**

3. 90° clockwise about the origin

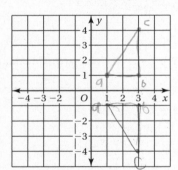

4. 270° counterclockwise about vertex A

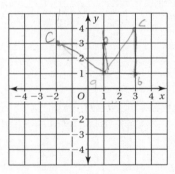

5. A triangle is rotated 180° about the origin. Its image is reflected in the x-axis. The vertices of the final triangle are $(-4, -4)$, $(-2, -4)$, and $(-3, -1)$. What are the vertices of the original triangle?

$(4, 4)$, $(2, 4)$ and $(3, 1)$

2.5 Similar Figures
For use with Activity 2.5

Essential Question How can you use proportions to help make decisions in art, design, and magazine layouts?

Original Photograph

In a computer art program, when you click and drag on a side of a photograph, you distort it.

But when you click and drag on a corner of the photograph, the dimensions remain proportional to the original.

Distorted

Distorted

Proportional

1 ACTIVITY: Reducing Photographs

Work with a partner. You are trying to reduce the photograph to the indicated size for a nature magazine. Can you reduce the photograph to the indicated size without distorting or cropping? Explain your reasoning.

a.

5 in.

6 in.

4 in.

5 in.

b.

6 in.

8 in.

3 in.

4 in.

2.5 **Similar Figures** (continued)

2 **ACTIVITY:** Creating Designs

Work with a partner.

a. Tell whether the dimensions of the new designs are proportional to the dimensions of the original design. Explain your reasoning.

Original

8   8

7

Design 1

7   7

6

Design 2

$6\frac{6}{7}$ $6\frac{6}{7}$

6

b. Draw two designs whose dimensions are proportional to the given design. Make one bigger and one smaller. Label the sides of the designs with their lengths.

5

4

8   10

6

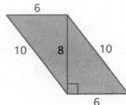

6

10   8   10

6

2.5 **Similar Figures** (continued)

What Is Your Answer?

3. **IN YOUR OWN WORDS** How can you use proportions to help make decisions in art, design, and magazine layouts? Give two examples.

4. **a.** Use a computer art program to draw two rectangles whose dimensions are proportional to each other.

   b. Print the two rectangles on the same piece of paper.

   c. Use a centimeter ruler to measure the length and the width of each rectangle. Record your measurements here.

"I love this statue. It seems similar to a big statue I saw in New York."

   d. Find the following ratios. What can you conclude?

   $$\frac{\text{Length of larger}}{\text{Length of smaller}} \qquad \frac{\text{Width of larger}}{\text{Width of smaller}}$$

Name _____ Date _____

Tell whether the two figures are similar. Explain your reasoning.

1.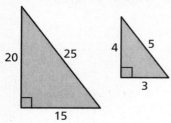

2.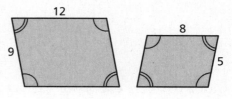

3. In your classroom, a dry erase board is 8 feet long and 4 feet wide. Your teacher makes individual dry erase boards for you to use at your desk that are 11.5 inches long and 9.5 inches wide. Are the boards similar?

4. You have a 4 x 6 photo of you and your friend.

   a. You order a 5 x 7 print of the photo. Is the new photo similar to the original?

   b. You enlarge the original photo to three times its size on your computer. Is the new photo similar to the original?

Perimeters and Areas of Similar Figures
For use with Activity 2.6

Essential Question How do changes in dimensions of similar geometric figures affect the perimeters and the areas of the figures?

---

1 ACTIVITY: Creating Similar Figures

Work with a partner. Use pattern blocks to make a figure whose dimensions are 2, 3, and 4 times greater than those of the original figure.*

  a. Square

  b. Rectangle

---

2 ACTIVITY: Finding Patterns for Perimeters

Work with a partner. Complete the table for the perimeter _P_ of each figure in Activity 1. Describe the pattern.

Figure	Original Side Lengths	Double Side Lengths	Triple Side Lengths	Quadruple Side Lengths
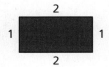	$P = $ _____			
	$P = $ _____			

*Cut-outs are available in the back of the Record and Practice Journal.

2.6 **Perimeters and Areas of Similar Figures** (continued)

3 **ACTIVITY:** Finding Patterns for Areas

Work with a partner. Complete the table for the area _A_ of each figure in Activity 1. Describe a pattern.

Figure	Original Side Lengths	Double Side Lengths	Triple Side Lengths	Quadruple Side Lengths
	$A = $ _____			
	$A = $ _____			

4 **ACTIVITY:** Drawing and Labeling Similar Figures

Work with a partner.

a. Find another rectangle that is similar and has one side from $(-1, -6)$ to $(5, -6)$. Label the vertices.

Check that the two rectangles are similar by showing that the ratios of corresponding sides are equal.

$$\frac{\text{Shaded Length}}{\text{Unshaded Length}} \overset{?}{=} \frac{\text{Shaded Width}}{\text{Unshaded Width}}$$

$$\frac{\text{change in } y}{\text{change in } y} \overset{?}{=} \frac{\text{change in } x}{\text{change in } x}$$

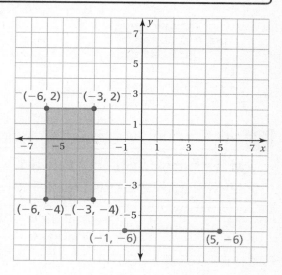

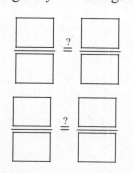

The ratios are _____ . So, the rectangles are _____ .

2.6 **Perimeters and Areas of Similar Figures** (continued)

b. Compare the perimeters and the areas of the figures. Are the results the same as your results from Activities 2 and 3? Explain.

c. There are three other rectangles that are similar to the shaded rectangle and have the given side.

• Draw each one. Label the vertices of each.

• Show that each is similar to the original shaded rectangle.

What Is Your Answer?

5. **IN YOUR OWN WORDS** How do changes in dimensions of similar geometric figures affect the perimeters and the areas of the figures?

6. What information do you need to know to find the dimensions of a figure that is similar to another figure? Give examples to support your explanation.

2.6 Practice
For use after Lesson 2.6

The two figures are similar. Find the ratios (shaded to nonshaded) of the perimeters and of the areas.

1.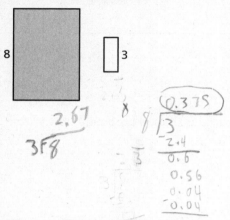

8

3

2,67
3√8

$\overset{(0.375)}{8\sqrt{3}}$
-2.4
0.6
0.56
0.04
0.04
0

2.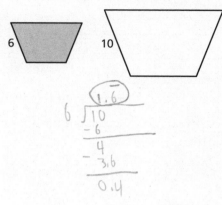

6

10

$\overset{1.\overline{6}}{6\sqrt{10}}$
-6
4
-3.6
0.4

The polygons are similar. Find x.

3.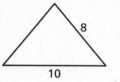

8

3

10

$x = 3.\overline{6}$

$\frac{x8}{10} \times \frac{3}{x}$ 30

$30 \div 8 = 3.\overline{6}$

$\dfrac{8}{3} = \dfrac{10}{x}$

4.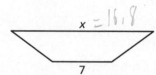

x = 16.8

12

7

5

$\frac{7}{x} \diagdown \frac{5}{12}$ $84 \div 5 = 16.8$

5. You buy two picture frames that are similar. The ratio of the corresponding side lengths is 4 : 5. What is the ratio of the areas?

$\overset{(1.25)}{4\sqrt{5}}$
-4
1
-0.8
0.12
0.05

4 : 5   .8 × larger

$\overset{0.8}{5\sqrt{4}}$

Name_____ Date_____

Essential Question How can you enlarge or reduce a figure in the coordinate plane?

1 **ACTIVITY:** Comparing Triangles in a Coordinate Plane

Work with a partner. Write the coordinates of the vertices of the shaded triangle. Then write the coordinates of the vertices of the nonshaded triangle.

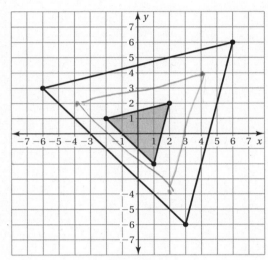

a. How are the two sets of coordinates related?

The coordinates are just multiplied by 3

b. How are the two triangles related? Explain your reasoning.

The triangles are the same but the clear one is dilated.

c. Draw a dashed triangle whose coordinates are twice the values of the corresponding coordinates of the shaded triangle. How are the dashed and shaded triangles related? Explain your reasoning.

The shaded triangle shows the same image of the other triangles just dilated.

2.7 **Dilations** (continued)

d. How are the coordinates of the nonshaded and dashed triangles related? How are the two triangles related? Explain your reasoning.

The coordinates have been multiplied from the shaded triangle.

2 **ACTIVITY:** Drawing Triangles in a Coordinate Plane

Work with a partner.

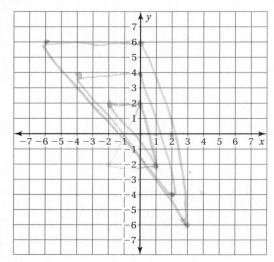

a. Draw the triangle whose vertices are $(0, 2), (-2, 2),$ and $(1, -2).$

b. Multiply each coordinate of the vertices by 2 to obtain three new vertices. Draw the triangle given by the three new vertices. How are the two triangles related?

The shape is the same but the sizes are different.

c. Repeat part (b) by multiplying by 3 instead of 2.

2.7 Dilations (continued)

3 **ACTIVITY:** Summarizing Transformations

Work with a partner. Make a table that summarizes the relationships between the original figure and its image for the four types of transformations you studied in this chapter.

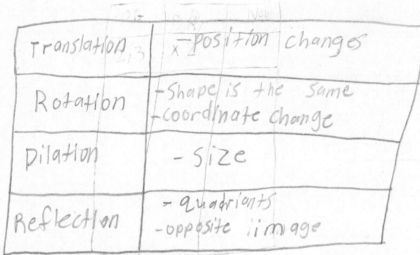

Translation	-position changes
Rotation	-Shape is the same -Coordinate change
Dilation	- size
Reflection	- quadrients -opposite iimage

What Is Your Answer?

4. **IN YOUR OWN WORDS** How can you enlarge or reduce a figure in the coordinate plane?

5. Describe how knowing how to enlarge or reduce figures in a technical drawing is important in a career such as drafting.

2.7 **Practice**
For use after Lesson 2.7

Tell whether the shaded figure is a dilation of the nonshaded figure.

1.

Yes

2.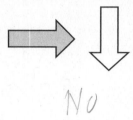

No

3.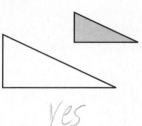

Yes

The vertices of a figure are given. Draw the figure and its image after a dilation with the given scale factor. Identify the type of dilation.

4. $A(-2, 2)$, $B(1, 2)$, $C(1, -1)$; $k = 3$

5. $D(4, 2)$, $E(4, 8)$, $F(8, 8)$, $G(8, 2)$; $k = \dfrac{1}{2}$

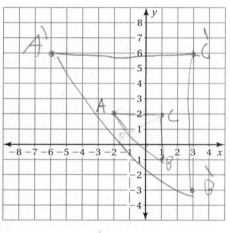

Enlargement,

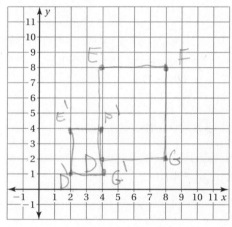

reduction.

6. A rectangle is dilated using a scale factor of 6. The image is then dilated using a scale factor of $\dfrac{1}{3}$. What scale factor could you use to dilate the original rectangle to get the final rectangle? Explain.

you could dilate the image using a scale factor of $\frac{1}{2}$.

because $4 \times 6 = 24 \times \frac{1}{3} = 8 \times \frac{1}{2} = 4$

$2 \times 6 = 12 \times \frac{1}{3} = 4 \times \frac{1}{2} = 2$

Name_____ Date _____

Tell whether the angles are *adjacent* or *vertical*. Then find the value of *x*.

1.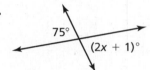
$x°$
$128°$

2.
$x°$
$35°$

3.
$75°$
$(2x + 1)°$

4.
$4x°$
$2x°$

5. The tree is tilted $14°$. Find the value of x.

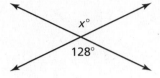
$14°$
$x°$

Chapter 3 **Fair Game Review** (continued)

Tell whether the angles are *complementary* or *supplementary*. Then find the value of x.

6.

7.
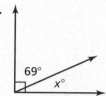

8.
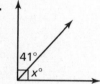

9.
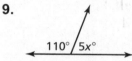

10. A tributary joins a river at an angle. Find the value of x.

3.1 Parallel Lines and Transversals
For use with Activity 3.1

Essential Question How can you describe angles formed by parallel lines and transversals?

---

1 ACTIVITY: A Property of Parallel Lines

Work with a partner.

- **Discuss what it means for two lines to be parallel. Decide on a strategy for drawing two parallel lines. Then draw the two parallel lines.**

- **Draw a third line that intersects the two parallel lines. This line is called a *transversal*.**

a. How many angles are formed by the parallel lines and the transversal? Label the angles.

b. Which of these angles have equal measures? Explain your reasoning.

3.1 **Parallel Lines and Transversals** (continued)

2 **ACTIVITY:** Creating Parallel Lines

Work with a partner.

a. If you were building the house in the photograph, how could you make sure that the studs are parallel to each other?

Studs

b. Identify sets of parallel lines and transversals in the photograph.

3 **ACTIVITY:** Using Technology to Draw Parallel Lines and a Transversal

Work with a partner. Use geometry software to draw two parallel lines intersected by a transversal.

a. Find all of the angle measures.

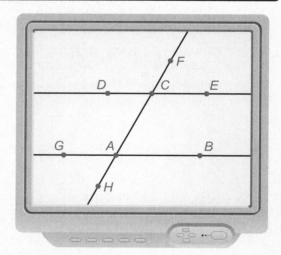

3.1 Parallel Lines and Transversals (continued)

b. Adjust the figure by moving the parallel lines or the transversal to a different position. Describe how the angle measures and relationships change.

What Is Your Answer?

4. IN YOUR OWN WORDS How can you describe angles formed by parallel lines and transversals? Give an example.

5. Use geometry software to draw a transversal that is perpendicular to two parallel lines. What do you notice about the angles formed by the parallel lines and the transversal?

3.1 **Practice**
For use after Lesson 3.1

Use the figure to find the measures of the numbered angles.

1.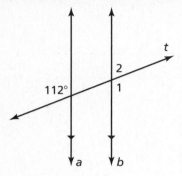

2.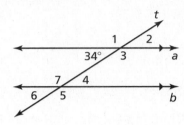

Complete the statement. Explain your reasoning.

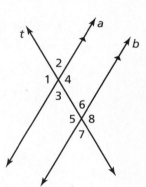

3. If the measure of ∠1 = 150°, then the measure of ∠6 = _____.

4. If the measure of ∠3 = 42°, then the measure of ∠5 = _____.

5. If the measure of ∠6 = 28°, then the measure of ∠3 = _____.

6. You paint a border around the top of the walls in your room. What angle does x need to be to repeat the pattern?

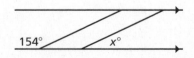

3.2 Angles of Triangles
For use with Activity 3.2

Essential Question How can you describe the relationships among the angles of a triangle?

> **1 ACTIVITY:** Exploring the Interior Angles of a Triangle

Work with a partner.

 a. Draw a triangle. Label the interior angles A, B, and C.

 b. Carefully cut out the triangle. Tear off the three corners of the triangle.

 c. Arrange angles A and B so that they share a vertex and are adjacent.

 d. How can you place the third angle to determine the sum of the measures of the interior angles? What is the sum?

 e. Compare your results with others in your class.

 f. **STRUCTURE** How does your result in part (d) compare to the rule you wrote in Lesson 1.1, Activity 2?

3.2 **Angles of Triangles** (continued)

2 **ACTIVITY:** Exploring the Interior Angles of a Triangle

Work with a partner.

a. Describe the figure.

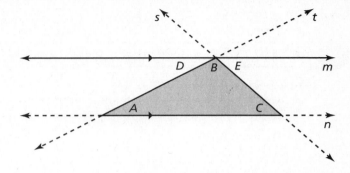

b. **LOGIC** Use what you know about parallel lines and transversals to justify your result in part (d) of Activity 1.

3 **ACTIVITY:** Exploring an Exterior Angle of a Triangle

Work with a partner.

a. Draw a triangle. Label the interior angles *A*, *B*, and *C*.

b. Carefully cut out the triangle.

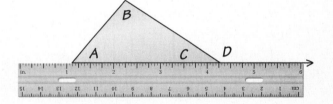

c. Place the triangle on a piece of paper and extend one side to form *exterior angle D*, as shown.

d. Tear off the corners that are not adjacent to the exterior angle. Arrange them to fill the exterior angle, as shown. What does this tell you about the measure of exterior angle *D*?

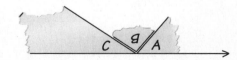

3.2 **Angles of Triangles** (continued)

4 **ACTIVITY:** Measuring the Exterior Angles of a Triangle

Work with a partner.

a. Draw a triangle and label the interior and exterior angles as shown.

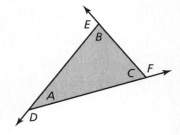

b. Use a protractor to measure all six angles of your triangle. Complete the table to organize your results. What does the table tell you about the measure of an exterior angle of a triangle?

Exterior Angle	$D = $ _____ °	$E = $ _____ °	$F = $ _____ °
Interior Angle	$B = $ _____ °	$A = $ _____ °	$A = $ _____ °
Interior Angle	$C = $ _____ °	$C = $ _____ °	$B = $ _____ °

What Is Your Answer?

5. **REPEATED REASONING** Draw three triangles that have different shapes. Repeat parts (b)–(d) from Activity 1 for each triangle. Do you get the same results? Explain.

6. **IN YOUR OWN WORDS** How can you describe the relationships among angles of a triangle?

3.2 Practice
For use after Lesson 3.2

Find the measures of the interior angles.

1.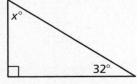

2.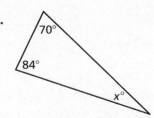

3.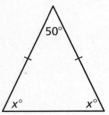

4.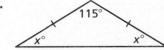

Find the measure of the exterior angle.

5.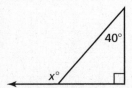

6.

7. Find the value of *x* on the clothes hanger.

3.3 Angles of Polygons
For use with Activity 3.3

Essential Question How can you find the sum of the interior angle measures and the sum of the exterior angle measures of a polygon?

1 | ACTIVITY: Exploring the Interior Angles of a Polygon

Work with a partner. In parts (a)–(e), identify each polygon and the number of sides *n*. Then find the sum of the interior angle measures of the polygon.

a. Polygon: _____   Number of sides: $n =$ _____

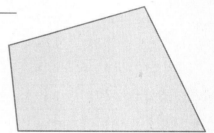

Draw a line segment on the figure that divides it into two triangles. Is there more than one way to do this? Explain.

What is the sum of the interior angle measures of each triangle?

What is the sum of the interior angle measures of the figure?

b.

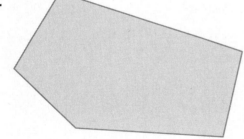

c.

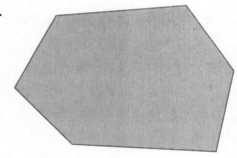

d.

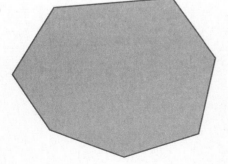

e.

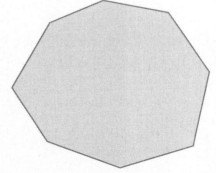

3.3 **Angles of Polygons** (continued)

f. REPEATED REASONING Use your results to complete the table. Then find the sum of the interior angle measures of a polygon with 12 sides.

Number of Sides, n	3	4	5	6	7	8
Number of Triangles						
Angle Sum, S						

2 **ACTIVITY:** Exploring the Exterior Angles of a Polygon

Work with a partner.

a. Draw a convex pentagon. Extend the sides to form the exterior angles. Label one exterior angle at each vertex A, B, C, D, and E, as shown.

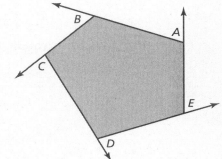

b. Cut out the exterior angles. How can you join the vertices to determine the sum of the angle measures? What do you notice?

c. REPEATED REASONING Repeat the procedure in parts (a) and (b) for each figure below.

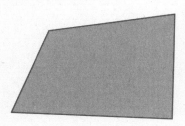
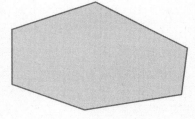

What can you conclude about the sum of the measures of the exterior angles of a convex polygon? Explain.

3.3 **Angles of Polygons** (continued)

What Is Your Answer?

3. **STRUCTURE** Use your results from Activity 1 to write an expression that represents the sum of the interior angle measures of a polygon.

4. **IN YOUR OWN WORDS** How can you find the sum of the interior angle measures and the sum of the exterior angle measures of a polygon?

3.3 Practice
For use after Lesson 3.3

Find the sum of the interior angle measures of the polygon.

1.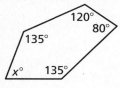

2.

3.

Find the measures of the interior angles.

4.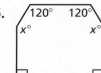
120° 80° 135° $x°$ 135°

5.
120° 120° $x°$ $x°$

Find the measure of each interior angle of the regular polygon.

6.

7.

8. In pottery class, you are making a pot that is shaped as a regular hexagon. What is the measure of each angle in the regular hexagon?

3.4 Using Similar Triangles
For use with Activity 3.4

Essential Question   How can you use angles to tell whether triangles are similar?

---

1   ACTIVITY: Constructing Similar Triangles

Work with a partner.

- **Use a straightedge to draw a line segment that is 4 centimeters long.**

- **Then use the line segment and a protractor to draw a triangle that has a 60° and a 40° angle as shown. Label the triangle** *ABC.*

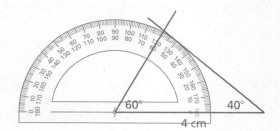

60°   40°

4 cm

a. Explain how to draw a larger triangle that has the same two angle measures. Label the triangle *JKL.*

b. Explain how to draw a smaller triangle that has the same two angle measures. Label the triangle *PQR.*

c. Are all of the triangles similar? Explain.

3.4 Using Similar Triangles (continued)

2 ACTIVITY: Using Technology to Explore Triangles

Work with a partner. Use geometry software to draw the triangle shown.

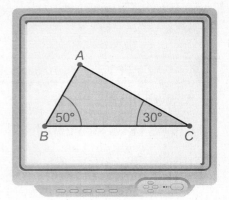

a. Dilate the triangle by the following scale factors.

2 $\dfrac{1}{2}$ $\dfrac{1}{4}$ 2.5

b. Measure the third angle in each triangle. What do you notice?

c. REASONING When two angles in one triangle are congruent to two angles in another triangle, can you conclude that the triangles are similar? Explain.

3 ACTIVITY: Indirect Measurement

Work with a partner.

a. Use the fact that two rays from the Sun are parallel to explain why $\triangle ABC$ and $\triangle DEF$ are similar.

x ft

Sun's ray

Sun's ray

5 ft

A 3 ft B

D 36 ft E

3.4 **Using Similar Triangles** (continued)

b. Explain how to use similar triangles to find the height of the flagpole.

What Is Your Answer?

4. IN YOUR OWN WORDS How can you use angles to tell whether triangles are similar?

5. PROJECT Work with a partner or in a small group.

a. Explain why the process in Activity 3 is called "indirect" measurement.

b. CHOOSE TOOLS Use indirect measurement to measure the height of something outside your school (a tree, a building, a flagpole). Before going outside, decide what materials you need to take with you.

c. MODELING Draw a diagram of the indirect measurement process you used. In the diagram, label the lengths that you actually measured and also the lengths that you calculated.

6. PRECISION Look back at Exercise 17 in Section 2.5. Explain how you can show that the two triangles are similar.

3.4 **Practice**
For use after Lesson 3.4

Tell whether the triangles are similar. Explain.

1.

2.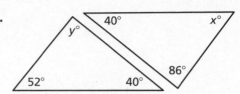

3.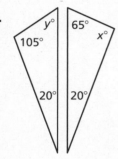

4.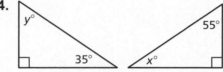

5. You can use similar triangles to find the
   height of a tree. Triangle *ABC* is similar
   to triangle *DEC*. What is the height of
   the tree?

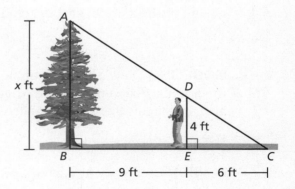

Chapter 4 Fair Game Review

Evaluate the expression when $x = \dfrac{1}{2}$ **and** $y = -5$.

1. $-2xy$

2. $4x^2 - 3y$

3. $\dfrac{10y}{12x + 4}$

4. $11x - 8(x - y)$

Evaluate the expression when $a = -9$ **and** $b = -4$.

5. $3ab$

6. $a^2 - 2(b + 12)$

7. $\dfrac{4b^2}{3b - 7}$

8. $7b^2 + 5(ab - 6)$

9. You go to the movies with five friends. You and one of your friends each buy a ticket and a bag of popcorn. The rest of your friends buy just one ticket each. The expression $4x + 2(x + y)$ represents the situation. Evaluate the expression when tickets cost \$7.25 and a bag of popcorn costs \$3.25.

Name _____   Date _____

Use the graph to answer the question.

10. Write the ordered pair that corresponds to Point *D*.

11. Write the ordered pair that corresponds to Point *H*.

12. Which point is located at $(-2, 4)$?

13. Which point is located at $(0, 3)$?

14. Which point(s) are located in Quadrant IV?

15. Which point(s) are located in Quadrant III?

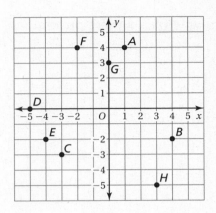

Plot the point.

16. $(3, -1)$

17. $(0, 2)$

18. $(-5, -4)$

19. $(-1, 0)$

20. $(-2, 3)$

4.1 Graphing Linear Equations
For use with Activity 4.1

Essential Question How can you recognize a linear equation? How can you draw its graph?

1 ACTIVITY: Graphing a Linear Equation

Work with a partner.

a. Use the equation $y = \frac{1}{2}x + 1$ to complete the table. (Choose any two x-values and find the y-values).

	Solution Points	
x		
$y = \frac{1}{2}x + 1$		

b. Write the two ordered pairs given by the table. These are called *solution points* of the equation.

c. PRECISION Plot the two solution points. Draw a line *exactly* through the two points.

d. Find a different point on the line. Check that this point is a solution point of the equation $y = \frac{1}{2}x + 1$.

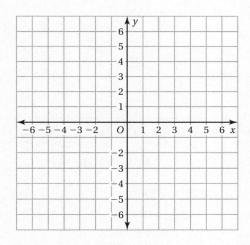

e. LOGIC Do you think it is true that *any* point on the line is a solution point of the equation $y = \frac{1}{2}x + 1$? Explain.

4.1 **Graphing Linear Equations** (continued)

f. Choose five additional *x*-values for the table. (Choose positive and negative *x*-values.) Plot the five corresponding solution points on the previous page. Does each point lie on the line?

	Solution Points				
x					
$y = \dfrac{1}{2}x + 1$					

g. **LOGIC** Do you think it is true that *any* solution point of the equation $y = \dfrac{1}{2}x + 1$ is a point on the line? Explain.

h. Why do you think $y = ax + b$ is called a *linear equation*?

2 **ACTIVITY:** Using a Graphing Calculator

Use a graphing calculator to graph $y = 2x + 5.$

a. Enter the equation $y = 2x + 5$ into your calculator.

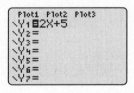

```
Plot1 Plot2 Plot3
\Y1■2X+5
\Y2=
\Y3=
\Y4=
\Y5=
\Y6=
\Y7=
```

b. Check the settings of the *viewing window*. The boundaries of the graph are set by the minimum and maximum *x*- and *y*-values. The numbers of units between the tick marks are set by the *x*- and *y*-scales.

```
WINDOW
Xmin=-10
Xmax=10
Xscl=1
Ymin=-10
Ymax=10
Yscl=1
Xres=1
```

This is the standard viewing window.

4.1 **Graphing Linear Equations** (continued)

c. Graph $y = 2x + 5$ on your calculator.

d. Change the settings of the viewing window to match those shown. Compare the two graphs.

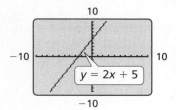

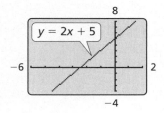

What Is Your Answer?

3. **IN YOUR OWN WORDS** How can you recognize a linear equation? How can you draw its graph? Write an equation that is linear. Write an equation that is *not* linear.

4. Use a graphing calculator to graph $y = 5x - 12$ in the standard viewing window.

   a. Can you tell where the line crosses the x-axis? Can you tell where the line crosses the y-axis?

   b. How can you adjust the viewing window so that you can determine where the line crosses the x- and y-axes?

5. **CHOOSE TOOLS** You want to graph $y = 2.5x - 3.8$. Would you graph it by hand or by using a graphing calculator? Why?

4.1 **Practice**
For use after Lesson 4.1

Graph the linear equation. Use a graphing calculator to check your graph, if possible.

1. $y = 4$

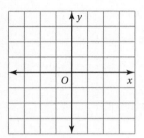

2. $y = -\dfrac{1}{3}x$

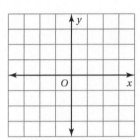

Solve for y. Then graph the equation. Use a graphing calculator to check your graph.

3. $y + 2x = 3$

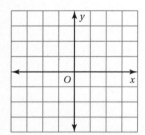

4. $2y - 3x = 1$

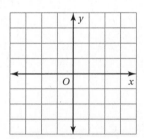

5. The equation $y = 2x + 4$ represents the cost y (in dollars) of renting a movie after x days of late charges.

a. Graph the equation.

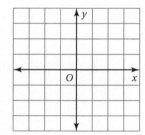

b. Use the graph to determine how much it costs after 3 days of late charges.

4.2 **Slope of a Line**
For use with Activity 4.2

Essential Question   How can the slope of a line be used to describe the line?

Slope is the rate of change between any two points on a line. It is the measure of the *steepness* of the line.

To find the slope of a line, find the ratio of the change in y (vertical change) to the change in x (horizontal change).

$$\text{slope} = \frac{\text{change in } y}{\text{change in } x}$$

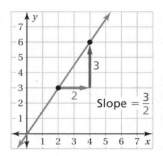

1   **ACTIVITY:** Finding the Slope of a Line

Work with a partner. Find the slope of each line using two methods.

   Method 1: Use the two black points.

   Method 2: Use the two gray points.

Do you get the same slope using each method? Why do you think this happens?

a.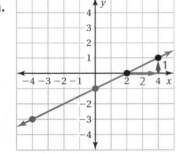

b.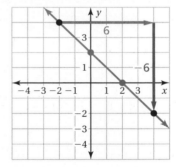

c.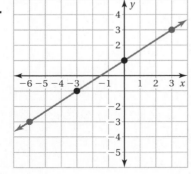

d.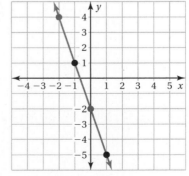

4.2 **Slope of a Line** (continued)

2 **ACTIVITY:** Using Similar Triangles

Work with a partner. Use the figure shown.

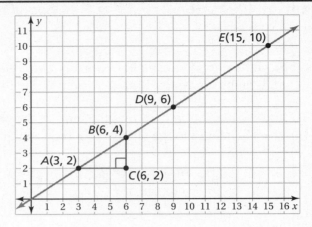

a. $\triangle ABC$ is a right triangle formed by drawing a horizontal line segment from point A and a vertical line segment from point B. Use this method to draw another right triangle, $\triangle DEF$.

b. What can you conclude about $\triangle ABC$ and $\triangle DEF$? Justify your conclusion.

c. For each triangle, find the ratio of the length of the vertical side to the length of the horizontal side. What do these ratios represent?

d. What can you conclude about the slope between any two points on the line?

3 **ACTIVITY:** Drawing Lines with Given Slopes

Work with a partner.

a. Draw two lines with slope $\frac{3}{4}$. One line passes through $(-4, 1)$, and the other line passes through $(4, 0)$. What do you notice about the two lines?

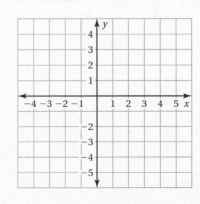

4.2 **Slope of a Line** (continued)

b. Draw two lines with slope $-\dfrac{4}{3}$. One line passes through $(2, 1)$, and the other line passes through $(-1, -1)$. What do you notice about the two lines?

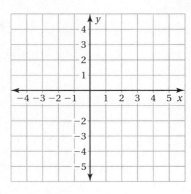

c. **CONJECTURE** Make a conjecture about two different nonvertical lines in the same plane that have the same slope.

d. Graph one line from part (a) and one line from part (b) in the same coordinate plane. Describe the angle formed by the two lines. What do you notice about the product of the slopes of the two lines?

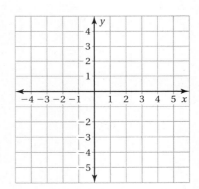

e. **REPEATED REASONING** Repeat part (d) for the two lines you did *not* choose. Based on your results, make a conjecture about two lines in the same plane whose slopes have a product of −1.

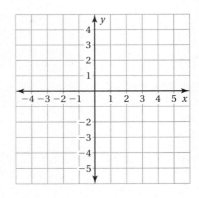

What Is Your Answer?

4. **IN YOUR OWN WORDS** How can you use the slope of a line to describe the line?

4.2 Practice
For use after Lesson 4.2

Find the slope of the line.

1.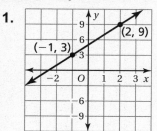

2.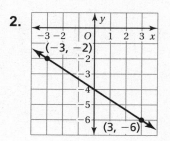

3.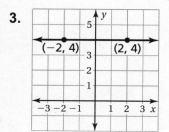

4.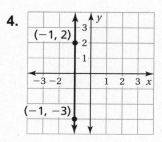

5. Which set of stairs is more difficult to climb? Explain.

10 in.

6 in.

Staircase 1

12 in.

8 in.

Staircase 2

Extension 4.2 **Practice**
For use after Extension 4.2

Which lines are parallel? How do you know?

1.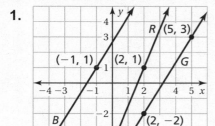

2.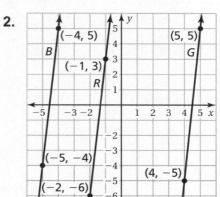

Are the given lines parallel? Explain your reasoning.

3. $y = 2, y = -4$

4. $x = 3, y = -3$

5. Is the quadrilateral a parallelogram? Justify your answer.

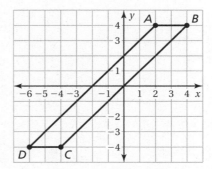

Name _____ Date _____

Which lines are perpendicular? How do you know?

6.
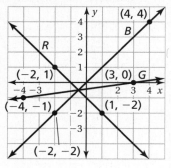

7.
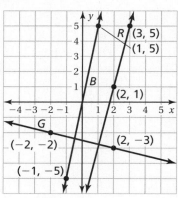

Are the given lines perpendicular? Explain your reasoning.

8. $x = 0, y = 3$

9. $y = 2, y = -\dfrac{1}{2}$

10. Is the parallelogram a rectangle? Justify your answer.

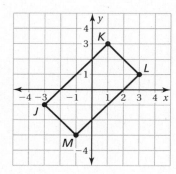

4.3 **Graphing Proportional Relationships**
For use with Activity 4.3

Essential Question How can you describe the graph of the equation $y = mx$?

1 **ACTIVITY:** Identifying Proportional Relationships

Work with a partner. Tell whether x and y are in a proportional relationship.
Explain your reasoning.

a. **Money**

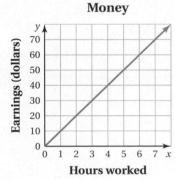

Hours worked

b. **Helicopter**

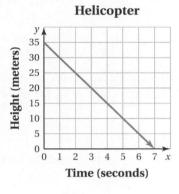

Time (seconds)

c. **Tickets**

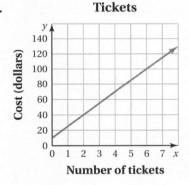

Number of tickets

d. **Pizzas**

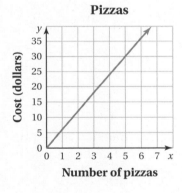

Number of pizzas

e.

Laps, x	1	2	3	4
Time (seconds), y	90	200	325	480

f.

Cups of Sugar, x	$\frac{1}{2}$	1	$1\frac{1}{2}$	2
Cups of Flour, y	1	2	3	4

4.3 Graphing Proportional Relationships (continued)

2 **ACTIVITY:** Analyzing Proportional Relationships

Work with a partner. Use only the proportional relationships in Activity 1 to do the following.

- **Find the slope of the line.**

- **Find the value of *y* for the ordered pair $(1, y)$.**

What do you notice? What does the value of *y* represent?

3 **ACTIVITY:** Deriving an Equation

Work with a partner. Let (x, y) represent any point on the graph of a proportional relationship.

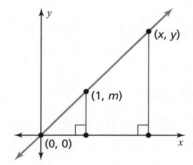

a. Explain why the two triangles are similar.

b. Because the triangles are similar, the corresponding side lengths are proportional. Use the vertical and horizontal side lengths to complete the steps below.

$$\frac{\boxed{}}{\boxed{}} = \frac{m}{1} \qquad \text{Ratios of side lengths}$$

$$\frac{\boxed{}}{\boxed{}} = m \qquad \text{Simplify.}$$

$$\boxed{} = m \bullet \boxed{} \qquad \text{Multiplication Property of Equality}$$

What does the final equation represent?

4.3 **Graphing Proportional Relationships** (continued)

c. Use your result in part (b) to write an equation that represents each proportional relationship in Activity 1.

What Is Your Answer?

4. **IN YOUR OWN WORDS** How can you describe the graph of the equation $y = mx$? How does the value of m affect the graph of the equation?

5. Give a real-life example of two quantities that are in a proportional relationship. Write an equation that represents the relationship and sketch its graph.

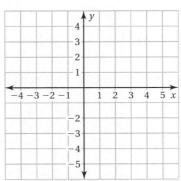

4.3 Practice
For use after Lesson 4.3

1. The amount p (in dollars) that you earn by working h hours is represented by the equation $p = 9h$. Graph the equation and interpret the slope.

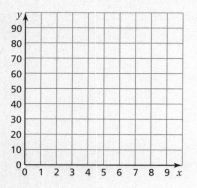

2. The cost c (in dollars) to rent a bicycle is proportional to the number h of hours that you rent the bicycle. It costs $20 to rent the bicycle for 4 hours.

   a. Write an equation that represents the situation.

   b. Interpret the slope.

   c. How much does it cost to rent the bicycle for 6 hours?

4.4 Graphing Linear Equations in Slope-Intercept Form
For use with Activity 4.4

Essential Question How can you describe the graph of the equation $y = mx + b$?

1 ACTIVITY: Analyzing Graphs of Lines

Work with a partner.

- Graph each equation.
- Find the slope of each line.
- Find the point where each line crosses the y-axis.
- Complete the table.

Equation	Slope of Graph	Point of Intersection with y-axis
a. $y = -\dfrac{1}{2}x + 1$		
b. $y = -x + 2$		
c. $y = -x - 2$		
d. $y = \dfrac{1}{2}x + 1$		
e. $y = x + 2$		
f. $y = x - 2$		
g. $y = \dfrac{1}{2}x - 1$		
h. $y = -\dfrac{1}{2}x - 1$		
i. $y = 3x + 2$		

4.4 **Graphing Linear Equations in Slope-Intercept Form** (continued)

Equation	Slope of Graph	Point of Intersection with *y*-axis
j. $y = 3x - 2$		

k. Do you notice any relationship between the slope of the graph and its equation? Between the point of intersection with the *y*-axis and its equation? Compare the results with those of other students in your class.

2 **ACTIVITY:** Deriving an Equation

Work with a partner.

a. Look at the graph of each equation in Activity 1. Do any of the graphs represent a proportional relationship? Explain.

b. For a nonproportional linear relationship, the graph crosses the *y*-axis at some point $(0, b)$, where b does not equal 0.

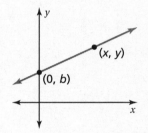

Let (x, y) represent any other point on the graph. You can use the formula for slope to write the equation for a nonproportional linear relationship.

Use the graph to complete the steps.

$$\frac{y_2 - y_1}{x_2 - x_1} = m$$ **Slope formula**

$$\frac{y - \boxed{}}{x - \boxed{}} = m$$ **Substitute values.**

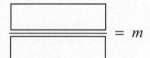 $= m$ **Simplify.**

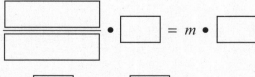 $\bullet \boxed{} = m \bullet \boxed{}$ **Multiplication Property of Equality**

$y - \boxed{} = m \bullet \boxed{}$ **Simplify.**

$y = m\boxed{} + \boxed{}$ **Addition Property of Equality**

4.4 **Graphing Linear Equations in Slope-Intercept Form** (continued)

c. What do m and b represent in the equation?

What Is Your Answer?

3. **IN YOUR OWN WORDS** How can you describe the graph of the equation $y = mx + b$?

   a. How does the value of m affect the graph of the equation?

   b. How does the value of b affect the graph of the equation?

   c. Check your answers to parts (a) and (b) with three equations that are not in Activity 1.

4. **LOGIC** Why do you think $y = mx + b$ is called the *slope-intercept form* of the equation of a line? Use drawings or diagrams to support your answer.

4.4 Practice
For use after Lesson 4.4

Find the slope and *y*-intercept of the graph of the linear equation.

1. $y = -3x + 9$

2. $y = 4 - \dfrac{2}{5}x$

3. $6 + y = 8x$

Graph the linear equation. Identify the *x*-intercept. Use a graphing calculator to check your answer.

4. $y = \dfrac{2}{3}x + 6$

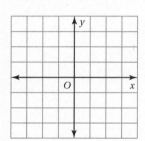

5. $y - 10 = -5x$

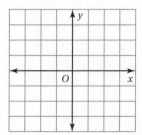

6. The equation $y = -90x + 1440$ represents the time (in minutes) left after x games of a tournament.

   a. Graph the equation.

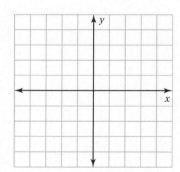

   b. Interpret the *x*-intercept and slope.

4.5 Graphing Linear Equations in Standard Form
For use with Activity 4.5

Essential Question How can you describe the graph of the equation $ax + by = c$?

---

1 **ACTIVITY:** Using a Table to Plot Points

Work with a partner. You sold a total of $16 worth of tickets to a school concert. You lost track of how many of each type of ticket you sold.

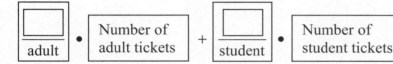

$$\boxed{} \cdot \frac{\text{Number of}}{\text{adult tickets}} + \boxed{} \cdot \frac{\text{Number of}}{\text{student tickets}} = \boxed{}$$

adult   student

a. Let x represent the number of adult tickets.
Let y represent the number of student tickets.
Write an equation that relates x and y.

b. Complete the table showing the different combinations of tickets you might have sold.

Number of Adult Tickets, x					
Number of Student Tickets, y					

c. Plot the points from the table. Describe the pattern formed by the points.

d. If you remember how many adult tickets you sold, can you determine how many student tickets you sold? Explain your reasoning.

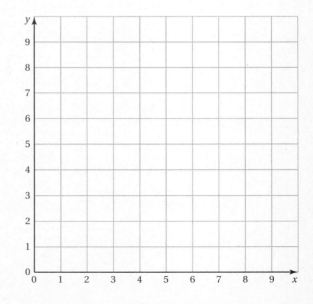

4.5 **Graphing Linear Equations in Standard Form** (continued)

2 **ACTIVITY:** Rewriting an Equation

Work with a partner. You sold a total of $16 worth of cheese. You forgot how many pounds of each type of cheese you sold.

CHEESE FOR SALE
Swiss: $4/lb   Cheddar: $2/lb

$$\boxed{} \cdot \boxed{\begin{array}{c}\text{Pounds}\\\text{of swiss}\end{array}} + \boxed{} \cdot \boxed{\begin{array}{c}\text{Pound of}\\\text{cheddar}\end{array}} = \boxed{}$$

$\quad$ lb $\qquad\qquad\qquad\qquad$ lb

a. Let x represent the number of pounds of swiss cheese.
Let y represent the number of pounds of cheddar cheese.
Write an equation that relates x and y.

b. Rewrite the equation in slope-intercept form. Then graph the equation.

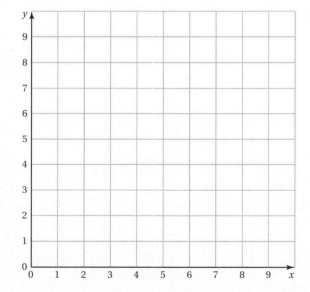

4.5 Graphing Linear Equations in Standard Form (continued)

c. You sold 2 pounds of cheddar cheese. How many pounds of swiss cheese did you sell?

d. Does the value $x = 2.5$ make sense in the context of the problem? Explain.

What Is Your Answer?

3. IN YOUR OWN WORDS How can you describe the graph of the equation $ax + by = c$?

4. Activities 1 and 2 show two different methods for graphing $ax + by = c$. Describe the two methods. Which method do you prefer? Explain.

5. Write a real-life problem that is similar to those shown in Activities 1 and 2.

6. Why do you think it might be easier to graph $x + y = 10$ without rewriting it in slope-intercept form and then graphing?

4.5 Practice
For use after Lesson 4.5

Write the linear equation in slope-intercept form.

1. $2x - y = 7$

2. $\frac{1}{4}x + y = -\frac{2}{7}$

3. $3x - 5y = -20$

Graph the linear equation using intercepts. Use a graphing calculator to check your graph.

4. $2x - 3y = 12$

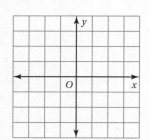

5. $x + 9y = -27$

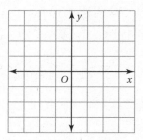

6. You go shopping and buy x shirts for $12 and y jeans for $28. The total spent is $84.

   a. Write an equation in standard form that models how much money you spent.

   b. Graph the equation and interpret the intercepts.

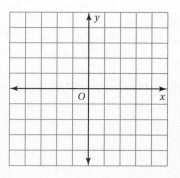

Name_____ Date _____

Essential Question How can you write an equation of a line when you are given the slope and *y*-intercept of the line?

1 **ACTIVITY:** Writing Equations of Lines

Work with a partner.

- **Find the slope of each line.**

- **Find the *y*-intercept of each line.**

- **Write an equation for each line.**

- **What do the three lines have in common?**

a.
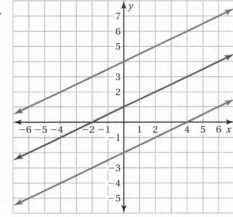

b.
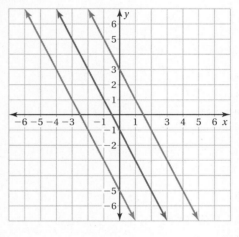

c.
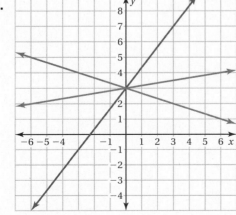

d.
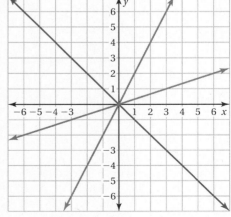

4.6 **Writing Equations in Slope-Intercept Form** (continued)

2 **ACTIVITY:** Describing a Parallelogram

Work with a partner.

- **Find the area of each parallelogram.**

- **Write an equation that represents each side of each parallelogram.**

a.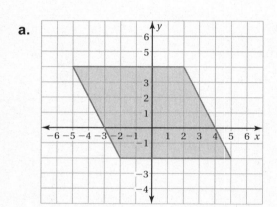

b.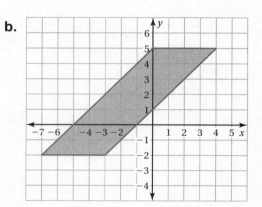

3 **ACTIVITY:** Interpreting the Slope and the *y*-Intercept

Work with a partner. The graph
shows a trip taken by a car, where
t is the time (in hours) and *y* is the
distance (in miles) from Phoenix.

a. Find the *y*-intercept of the graph.
What does it represent?

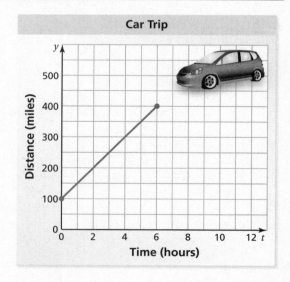

Car Trip

4.6 **Writing Equations in Slope-Intercept Form** (continued)

b. Find the slope of the graph. What does it represent?

c. How long did the trip last?

d. How far from Phoenix was the car at the end of the trip?

e. Write an equation that represents the graph.

What Is Your Answer?

4. IN YOUR OWN WORDS How can you write an equation of a line when you are given the slope and the y-intercept of the line? Give an example that is different from those in Activities 1, 2, and 3.

5. Two sides of a parallelogram are represented by the equations $y = 2x + 1$ and $y = -x + 3$. Give two equations that can represent the other two sides.

Name _____ Date _____

Write an equation of the line in slope-intercept form.

1.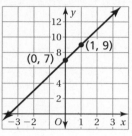

2.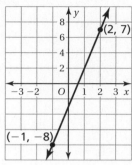

3.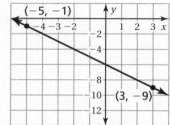

4.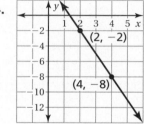

Write an equation of the line that passes through the points.

5. $(3, 8), (-2, 8)$

6. $(4, 3), (6, -3)$

7. $(-1, 0), (-5, 0)$

8. You organize a garage sale. You have $30 at the beginning of the sale. You earn an average of $20 per hour. Write an equation that represents the amount of money y you have after x hours.

Name_____ Date_____

4.7 Writing Equations in Point-Slope Form
For use with Activity 4.7

Essential Question How can you write an equation of a line when you are given the slope and a point on the line?

1 ACTIVITY: Writing Equations of Lines

Work with a partner.

- Sketch the line that has the given slope and passes through the given point.
- Find the *y*-intercept of the line.
- Write an equation of the line.

a. $m = -2$

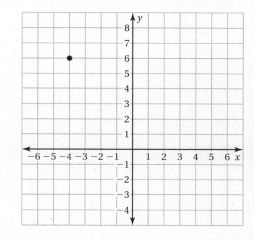

b. $m = \dfrac{1}{3}$

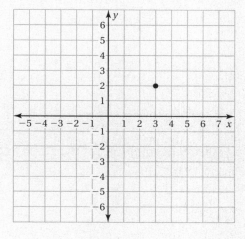

c. $m = -\dfrac{2}{3}$

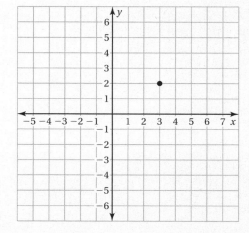

d. $m = \dfrac{5}{2}$

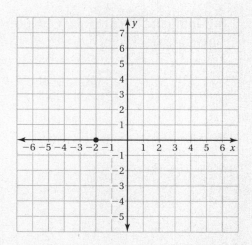

4.7 **Writing Equations in Point-Slope Form** (continued)

2 **ACTIVITY:** Deriving an Equation

Work with a partner.

a. Draw a nonvertical line that passes through the point (x_1, y_1).

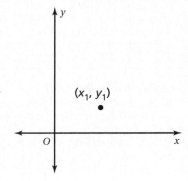

b. Plot another point on your line. Label this point as (x, y). This point represents any other point on the line.

c. Label the rise and run of the line through the points (x_1, y_1) and (x, y).

d. The rise can be written as $y - y_1$. The run can be written as $x - x_1$. Explain why this is true.

e. Write an equation for the slope m of the line using the expressions from part (d).

f. Multiply each side of the equation by the expression in the denominator. Write your result. What does this result represent?

4.7 Writing Equations in Point-Slope Form (continued)

3 ACTIVITY: Writing an Equation

Work with a partner.

For 4 months, you saved $25 a month. You now have $175 in your savings account.

- Draw a graph that shows the balance in your account after t months.

- Use your result from Activity 2 to write an equation that represents the balance A after t months.

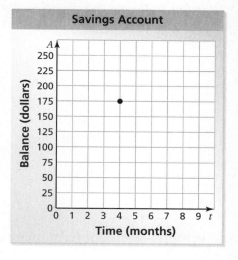

What Is Your Answer?

4. Redo Activity 1 using the equation you found in Activity 2. Compare the results. What do you notice?

5. Why do you think $y - y_1 = m(x - x_1)$ is called the *point-slope form* of the equation of a line? Why do you think this is important?

6. IN YOUR OWN WORDS How can you write an equation of a line when you are given the slope and a point on the line? Give an example that is different from those in Activity 1.

4.7 Practice
For use after Lesson 4.7

Write in point-slope form an equation of the line that passes through the given point that has the given slope.

1. $m = -3; (-4, 6)$

2. $m = -\dfrac{4}{3}; (3, -1)$

Write in slope-intercept form an equation of the line that passes through the given points.

3. $(-3, 0), (-2, 3)$

4. $(-6, 10), (6, -10)$

5. The total cost for bowling includes the fee for shoe rental plus a fee per game. The cost of each game increases the price by \$4. After 3 games, the total cost with shoe rental is \$14.

   a. Write an equation to represent the total cost y to rent shoes and bowl x games.

   b. How much is shoe rental? How is this represented in the equation?

Chapter 5 Fair Game Review

Simplify the expression.

1. $2x + 5 - x$

2. $4 + 2d - 4d$

3. $7y - 8 + 6y - 3$

4. $5 + 4z - 3 + 3z$

5. $4(s + 2) + s - 1$

6. $2(4x - 5) - 3$

7. The width of a garden is $(4x - 1)$ feet and the length is $2x$ feet. Find the perimeter of the garden.

Chapter 5 **Fair Game Review** (continued)

Solve the equation. Check your solution.

8. $8y - 3 = 13$

9. $4a + 11 - a = 2$

10. $9 = 4(3k - 4) - 7k$

11. $-12 - 5(6 - 2m) = 18$

12. $15 - t + 8t = -13$

13. $5h - 2\left(\dfrac{3}{2}h + 4\right) = 10$

14. The profit P (in dollars) from selling x calculators is $P = 25x - (10x + 250)$. How many calculators are sold when the profit is \$425?

5.1 Solving Systems of Linear Equations by Graphing
For use with Activity 5.1

Essential Question   How can you solve a system of linear equations?

1   ACTIVITY: Writing a System of Linear Equations

Work with a partner.

Your family starts a bed-and-breakfast. It spends $500 fixing up a bedroom to rent. The cost for food and utilities is $10 per night. Your family charges $60 per night to rent the bedroom.

a. Write an equation that represents the costs.

$$\boxed{\begin{array}{c}\text{Cost, } C \\ \text{(in dollars)}\end{array}} = \boxed{\begin{array}{c}\$10 \text{ per} \\ \text{night}\end{array}} \cdot \boxed{\begin{array}{c}\text{Number of} \\ \text{nights, } x\end{array}} + \boxed{\$500}$$

b. Write an equation that represents the revenue (income).

$$\boxed{\begin{array}{c}\text{Revenue, } R \\ \text{(in dollars)}\end{array}} = \boxed{\begin{array}{c}\$60 \text{ per} \\ \text{night}\end{array}} \cdot \boxed{\begin{array}{c}\text{Number of} \\ \text{nights, } x\end{array}}$$

c. A set of two (or more) linear equations is called a **system of linear equations**. Write the system of linear equations for this problem.

Name _____ Date _____

5.1 Solving Systems of Linear Equations by Graphing (continued)

2 ACTIVITY: Using a Table to Solve a System

Work with a partner. Use the cost and revenue equations from Activity 1 to find how many nights your family needs to rent the bedroom before recovering the cost of fixing up the bedroom. This is the *break-even point*.

a. Complete the table.

x	0	1	2	3	4	5	6	7	8	9	10	11
C												
R												

b. How many nights does your family need to rent the bedroom before breaking even?

3 ACTIVITY: Using a Graph to Solve a System

Work with a partner.

a. Graph the cost equation from Activity 1.

b. In the same coordinate plane, graph the revenue equation from Activity 1.

c. Find the point of intersection of the two graphs. What does this point represent? How does this compare to the break-even point in Activity 2? Explain.

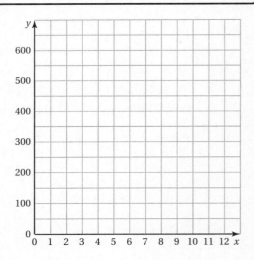

5.1 Solving Systems of Linear Equations by Graphing (continued)

4 ACTIVITY: Using a Graphing Calculator

Work with a partner. Use a graphing calculator to solve the system.

$y = 10x + 500$ **Equation 1**

$y = 60x$ **Equation 2**

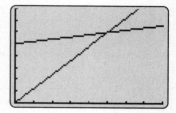

a. Enter the equations into your calculator. Then graph the equations. What is an appropriate window?

b. On your graph, how can you determine which line is the graph of which equation? Label the equations on the graph shown.

c. Visually estimate the point of intersection of the graphs.

d. To find the solution, use the *intersect* feature to find the point of intersection. The solution is at (_____ , _____).

What Is Your Answer?

5. **IN YOUR OWN WORDS** How can you solve a system of linear equations? How can you check your solution?

6. **CHOOSE TOOLS** Solve one of the systems by using a table, another system by sketching a graph, and the remaining system by using a graphing calculator. Explain why you chose each method.

a. $y = 4.3x + 1.2$

    $y = -1.7x - 2.4$

b. $y = x$

    $y = -2x + 9$

c. $y = -x - 5$

    $y = 3x + 1$

5.1 Practice
For use after Lesson 5.1

1. Use the table to find the break-even point. Check your solution.

   $C = 25x + 210$

   $R = 60x$

x	0	1	2	3	4	5	6	7	8
C									
R									

Solve the system of linear equations by graphing.

2. $y = 3x + 1$

   $y = -2x - 4$

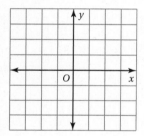

3. $y = -4x + 1$

   $y = 5x - 8$

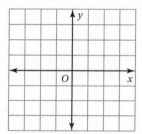

Use a graphing calculator to solve the system of linear equations.

4. $y = \dfrac{2}{3}x + 2$

   $x - y = 4$

5. $y = x - 7$

   $y + x = 3$

6. There are 26 students in your class. There are 4 more girls than boys. Use a system of linear equations to find how many boys are in your class. How many girls are in your class?

5.2 Solving Systems of Linear Equations by Substitution
For use with Activity 5.2

Essential Question How can you use substitution to solve a system of linear equations?

1 ACTIVITY: Using Substitution to Solve a System

Work with a partner. Solve each system of linear equations using two methods.

Method 1: Solve for x first.

Solve for x in one of the equations. Use the expression for x to find the solution of the system. Explain how you did it.

Method 2: Solve for y first.

Solve for y in one of the equations. Use the expression for y to find the solution of the system. Explain how you did it.

Is the solution the same using both methods?

a. $6x - y = 11$
$2x + 3y = 7$

b. $2x - 3y = -1$
$x - y = 1$

c. $3x + y = 5$
$5x - 4y = -3$

d. $5x - y = 2$
$3x - 6y = 12$

e. $x + y = -1$
$5x + y = -13$

f. $2x - 6y = -6$
$7x - 8y = 5$

5.2 **Solving Systems of Linear Equations by Substitution** (continued)

2 **ACTIVITY:** Writing and Solving a System of Equations

Work with a partner.

a. Roll a pair of number cubes that
have different colors. Then write the
ordered pair shown by the number cubes.
The ordered pair at the right is (3, 4).

x-value

y-value

b. Write a system of linear equations that
has this ordered pair as its solution.

c. Exchange systems with your partner and use one of the methods from
Activity 1 to solve the system.

3 **ACTIVITY:** Solving a Secret Code

Work with a partner. Decode the quote by Archimedes.

___ ___ ___ ___ ___ ___ ___ ___ ___ ___ ___ ___ ___ ___ ___ ___ ___ ___ ___ ,
−8 −7 7 −5 −4 −5 −3 −2 −1 −3 0 −5 1 2 3 1 −3 4 5

___ ___ ___ ___ ___ ___ ___ ___ ___ ___ ___ ___ ___ ___ ___ ___ ___ ___ ___ .
−3 4 5 −7 6 −7 −1 −1 −4 2 7 −5 1 8 −5 −5 −3 9 1 8

5.2 **Solving Systems of Linear Equations by Substitution** (continued)

$(\mathbf{A, C})$ $x + y = -3$ $(\mathbf{D, E})$ $x + y = 0$ $(\mathbf{G, H})$ $x + y = 0$
           $x - y = -3$ $x - y = 10$ $x - y = -16$

$(\mathbf{I, L})$ $x + 2y = -9$ $(\mathbf{M, N})$ $x + 2y = 4$ $(\mathbf{O, P})$ $x + 2y = -2$
          $2x - y = -13$ $2x - y = -12$ $2x - y = 6$

$(\mathbf{R, S})$ $2x + y = 21$ $(\mathbf{T, U})$ $2x + y = -7$ $(\mathbf{V, W})$ $2x + y = 20$
          $x - y = 6$ $x - y = 10$ $x - y = 1$

What Is Your Answer?

4. **IN YOUR OWN WORDS** How can you use substitution to solve a system of linear equations?

5.2 Practice

For use after Lesson 5.2

Solve the system of linear equations by substitution. Check your solution.

1. $y = -2x + 4$

   $-x + 3y = -9$

2. $\dfrac{3}{4}x - 5y = 7$

   $x = -4y + 12$

3. $5x - y = 4$

   $2x + 2y = 16$

4. $2x + 3y = 0$

   $8x + 9y = 18$

5. A gas station sells a total of 4500 gallons of regular gas and premium gas in one day. The ratio of gallons of regular gas sold to gallons of premium gas sold is $7 : 2$.

   a. Write a system of linear equations that represents this situation.

   b. How many gallons sold were regular gas? premium gas?

5.3 Solving Systems of Linear Equations by Elimination
For use with Activity 5.3

Essential Question How can you use elimination to solve a system of linear equations?

1   ACTIVITY: Using Elimination to Solve a System

Work with a partner. Solve each system of linear equations using two methods.

Method 1: Subtract.

Subtract Equation 2 from Equation 1. What is the result? Explain how you can use the result to solve the system of equations.

Method 2: Add.

Add the two equations. What is the result? Explain how you can use the result to solve the system of equations.

Is the solution the same using both methods?

   a. $2x + y = 4$ **b.** $3x - y = 4$ **c.** $x + 2y = 7$

       $2x - y = 0$ $3x + y = 2$ $x - 2y = -5$

2   ACTIVITY: Using Elimination to Solve a System

Work with a partner.

   $2x + y = 2$ Equation 1

   $x + 5y = 1$ Equation 2

   a. Can you add or subtract the equations to solve the system of linear equations? Explain.

5.3 **Solving Systems of Linear Equations by Elimination** (continued)

b. Explain what property you can apply to Equation 1 in the system so that the *y* coefficients are the same.

c. Explain what property you can apply to Equation 2 in the system so that the *x* coefficients are the same.

d. You solve the system in part (b). Your partner solves the system in part (c). Compare your solutions.

e. Use a graphing calculator to check your solution.

3 **ACTIVITY:** Solving a Secret Code

Work with a partner. Solve the puzzle to find the name of a famous mathematician who lived in Egypt around 350 A.D.

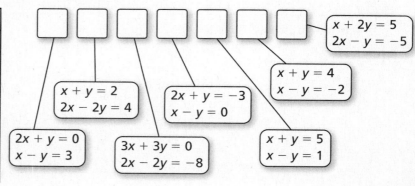

4	B	W	R	M	F	Y	K	N
3	O	J	A	S	I	D	X	Z
2	Q	P	C	E	G	B	T	J
1	M	R	C	Z	N	O	U	W
0	K	X	U	H	L	Y	S	Q
−1	F	E	A	S	W	K	R	M
−2	G	J	Z	N	H	V	D	G
−3	E	L	X	L	F	Q	O	B

−3 −2 −1 0 1 2 3 4

$x + 2y = 5$
$2x - y = -5$

$x + y = 4$
$x - y = -2$

$x + y = 2$
$2x - 2y = 4$

$2x + y = -3$
$x - y = 0$

$x + y = 5$
$x - y = 1$

$2x + y = 0$
$x - y = 3$

$3x + 3y = 0$
$2x - 2y = -8$

5.3 Solving Systems of Linear Equations by Elimination (continued)

What Is Your Answer?

4. **IN YOUR OWN WORDS** How can you use elimination to solve a system of linear equations?

5. **STRUCTURE** When can you add or subtract equations in a system to solve the system? When do you have to multiply first? Justify your answers with examples.

6. **LOGIC** In Activity 2, why can you multiply the equations in the system by a constant and not change the solution of the system? Explain your reasoning.

5.3 Practice
For use after Lesson 5.3

Solve the system of linear equations by elimination. Check your solution.

1. $x + y = 7$

$3x - y = 1$

2. $-2x - 5y = -8$

$-2x + y = 16$

3. $8x - 9y = 7$

$2x - 3y = -5$

4. $-5x + 3y = -6$

$9x - 4y = 1$

5. A high school has a total of 850 students. There are 60 more female students than there are male students.

    a. Write a system of linear equations that represents this situation.

    b. How many students are female? male?

Name_____ Date_____

Essential Question Can a system of linear equations have no solution? Can a system of linear equations have many solutions?

1 **ACTIVITY:** Writing a System of Linear Equations

Work with a partner. Your cousin is 3 years older than you. Your ages can be represented by two linear equations.

$y = t$ Your age

$y = t + 3$ Your cousin's age

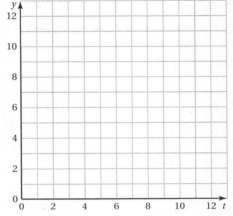

a. Graph both equations in the same coordinate plane.

b. What is the vertical distance between the two graphs? What does this distance represent?

c. Do the two graphs intersect? Explain what this means in terms of your age and your cousin's age.

2 **ACTIVITY:** Using a Table to Solve a System

Work with a partner. You invest $500 for equipment to make dog backpacks. Each backpack costs you $15 for materials. You sell each backpack for $15.

a. Complete the table for your cost C and your revenue R.

x	0	1	2	3	4	5	6	7	8	9	10
C											
R											

5.4 **Solving Special Systems of Linear Equations** (continued)

b. When will you break even? What is wrong?

3 **ACTIVITY:** Using a Graph to Solve a Puzzle

Work with a partner. Let x and y be two numbers. Here are two clues about the values of x and y.

	Words	**Equation**
Clue 1:	The value of y is 4 more than twice the value of x.	$y = 2x + 4$
Clue 2:	The difference of $3y$ and $6x$ is 12.	$3y - 6x = 12$

a. Graph both equations in the same coordinate plane.

b. Do the two lines intersect? Explain.

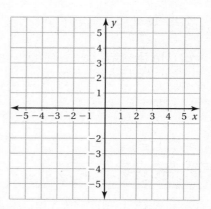

c. What is the solution of the puzzle?

d. Use the equation $y = 2x + 4$ to complete the table.

x	0	1	2	3	4	5	6	7	8	9	10
y											

5.4 Solving Special Systems of Linear Equations (continued)

e. Does each solution in the table satisfy *both* clues?

f. What can you conclude? How many solutions does the puzzle have? How can you describe them?

What Is Your Answer?

4. IN YOUR OWN WORDS Can a system of linear equations have no solution? Can a system of linear equations have many solutions? Give examples to support your answers.

5.4 Practice
For use after Lesson 5.4

Solve the system of linear equations. Check your solution.

1. $y = 2x - 5$

    $y = 2x + 7$

2. $3x + 4y = -10$

    $y = -\dfrac{3}{4}x - \dfrac{5}{2}$

3. $x - y = 8$

    $2y = 2x - 16$

4. $3y = -6x + 4$

    $2x + y = 9$

5. You start reading a book for your literature class two days before your friend. You both read 10 pages per night. A system of linear equations that represents this situation is $y = 10x + 20$ and $y = 10x$. Will your friend finish the book before you? Justify your answer.

Name_____ Date _____

Extension 5.4 **Practice**
For use after Extension 5.4

Use a graph to solve the equation. Check your solution.

1. $3x - 4 = -x$

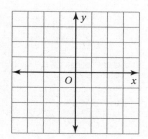

2. $\frac{1}{3}x + 3 = 4x - 8$

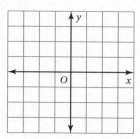

3. $\frac{1}{2}x + 4 = -x - 11$

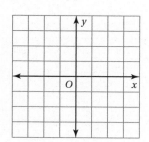

4. $-x + 1 = -\frac{1}{4}x - \frac{1}{2}$

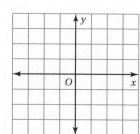

5. On the first day of your garage sale, you earned $12x + 9$ dollars. The next day you earned $22x$ dollars. Is it possible that you earned the same amount each day? Explain.

Extension 5.4 **Practice** (continued)

6. You hike uphill at a rate of 200 feet per minute. Your friend hikes downhill on the same trail at a rate of 250 feet per minute. How long will it be until you meet?

1800 ft
100 ft
Not drawn to scale

7. Two savings accounts earn simple interest. Account A has a beginning balance of $500 and grows by $25 per year. Account B has a beginning balance of $750 and grows by $15 per year.

| Growth rate | • | Years, x | + | Beginning balance | = | Growth rate | • | Years, x | + | Beginning balance |

a. Use the model to write an equation.

b. After how many years *x* do the accounts have the same balance?

Name_____ Date_____

Chapter 6

Fair Game Review

Find the missing value in the table.

1.

x	y
1	5
3	7
5	9
7	

2.

x	y
2	6
4	12
8	24
12	

3.

x	y
6	11
14	19
26	31
41	

4.

x	y
8	4
18	9
28	14
38	

5.

x	y
4	2.5
11	9.5
15	13.5
21	

6.

x	y
6	5.8
15	14.8
22.8	22.6
31.4	

Name _____ Date _____

Evaluate the expression when $x = 2$, $y = 3$, and $z = -4$.

7. $3x - 2$

8. $-6 - 2y$

9. $2z^2$

10. $3y - 3z$

11. $\dfrac{8}{x} - 1$

12. $-1 + \dfrac{z}{2}$

6.1 Relations and Functions
For use with Activity 6.1

Essential Question How can you use a mapping diagram to show the relationship between two data sets?

1 **ACTIVITY:** Constructing Mapping Diagrams

Work with a partner. Complete the mapping diagram.

a. Area A

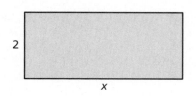

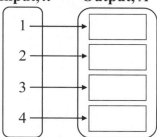

b. Perimeter P

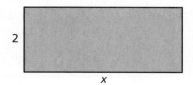

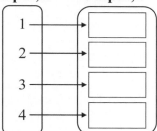

c. Circumference C

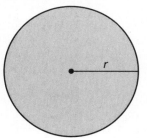

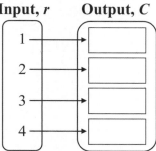

d. Volume V

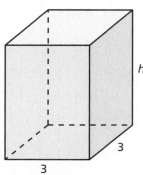

Name _____ Date _____

6.1   **Relations and Functions** (continued)

2   **ACTIVITY:** Describing Situations

Work with a partner. How many outputs are assigned to each input?
Describe a possible situation for each mapping diagram.

a.   **Input, *x***   **Output, *y***

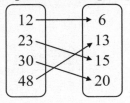

b.   **Input, *x***   **Output, *y***

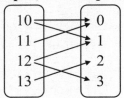

3   **ACTIVITY:** Interpreting Mapping Diagrams

Work with a partner. Describe the pattern in the mapping diagram.
Complete the diagram.

a.   **Input, *t***   **Output, *M***

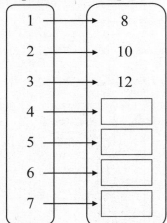

122 **Big Ideas Math Blue**
Record and Practice Journal

Copyright © Big Ideas Learning, LLC
All rights reserved.

6.1 Relations and Functions (continued)

b.

Input, x	Output, A
1	4/3
2	5/3
3	2
4	
5	
6	
7	

What Is Your Answer?

4. **IN YOUR OWN WORDS** How can you use a mapping diagram to show the relationship between two data sets?

"I made a mapping diagram."

"It shows how I feel about my skateboard with each passing day."

6.1 Practice
For use after Lesson 6.1

List the ordered pairs shown in the mapping diagram.

1. Input   Output

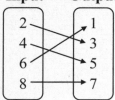

2. Input   Output

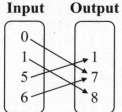

3. Draw a mapping diagram for the graph. Then describe the pattern of inputs and outputs.

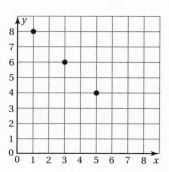

4. The table shows the number of beads needed to make a bracelet. Use the table to draw a mapping diagram.

Bracelet Length (in.)	Number of Beads
6	12
7	14
8	16
9	18

Name_____ Date_____

6.2 Representations of Functions

For use with Activity 6.2

Essential Question How can you represent a function in different ways?

---

1 **ACTIVITY:** Describing a Function

Work with a partner. Complete the mapping diagram on the next page for the area of the figure. Then write an equation that describes the function.

a.
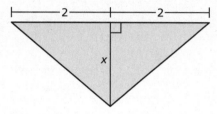

b.
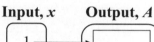
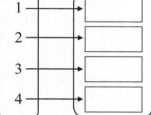

Input, *x*	Output, *A*	Input, *x*	Output, *A*
1 →		1 →	
2 →		2 →	
3 →		3 →	
4 →		4 →	

---

2 **ACTIVITY:** Using a Table

Work with a partner. Make a table that shows the pattern for the area, where the input is the figure number *x* and the output is the area *A*. Write an equation that describes the function. Then use your equation to find which figure has an area of 81 when the pattern continues.

1 square unit

a.

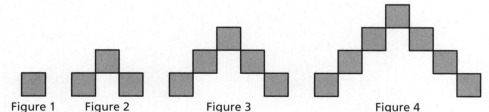

Figure 1   Figure 2   Figure 3   Figure 4

6.2 **Representations of Functions** (continued)

b.

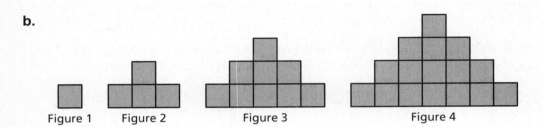

Figure 1   Figure 2   Figure 3   Figure 4

3 **ACTIVITY:** Using a Graph

Work with a partner. Graph the data. Use the graph to test the truth of each statement. If the statement is true, write an equation that shows how to obtain one measurement from the other measurement.

a. "You can find the horsepower of a race car engine if you know its volume in cubic inches."

Volume (cubic inches), x	200	350	350	500
Horsepower, y	375	650	250	600

Race Car Engine

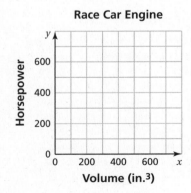

6.2 **Representations of Functions** (continued)

b. "You can find the volume of a race car engine in cubic centimeters if you know its volume in cubic inches."

Volume (cubic inches), *x*	100	200	300
Volume (cubic centimeters), *y*	1640	3280	4920

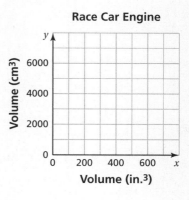

Race Car Engine

---

4 **ACTIVITY:** Interpreting a Graph

Work with a partner. The table shows the average speeds of the winners of the Daytona 500. Graph the data. Can you use the graph to predict future winning speeds? Explain why or why not.

Year	2004	2005	2006	2007	2008	2009	2010	2011	2012
Speed (mi/h)	156	135	143	149	153	133	137	130	140

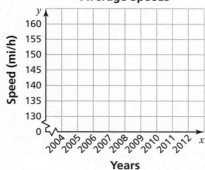

What Is Your Answer?

5. IN YOUR OWN WORDS How can you represent a function in different ways?

Name _____ Date _____

Practice
For use after Lesson 6.2

Write a function rule for the statement.

1. The output is four times the input.

2. The output is eight less than the input.

Find the value of *y* for the given value of *x*.

3. $y = \dfrac{x}{3}; x = 12$

4. $y = 5x + 9; x = 2$

5. You set up a hot chocolate stand at a football game. The cost of your supplies is $75. You charge $0.50 for each cup of hot chocolate.

   a. Write a function that represents the profit P for selling c cups of hot chocolate.

   b. You will *break even* when the cost of your supplies equals your income. How many cups of hot chocolate must you sell to break even?

Name_____ Date_____

6.3 Linear Functions
For use with Activity 6.3

Essential Question How can you use a function to describe a linear pattern?

1 ACTIVITY: Finding Linear Patterns

Work with a partner.

- Plot the points from the table in a coordinate plane.

- Write a linear equation for the function.

a.

x	0	2	4	6	8
y	150	125	100	75	50

b.

x	4	6	8	10	12
y	15	20	25	30	35

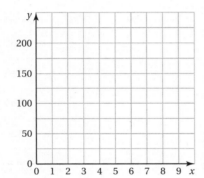

c.

x	−4	−2	0	2	4
y	4	6	8	10	12

d.

x	−4	−2	0	2	4
y	1	0	−1	−2	−3

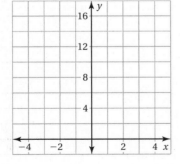

6.3   **Linear Functions** (continued)

2 **ACTIVITY:** Finding Linear Patterns

Work with a partner. The table shows a familiar linear pattern from geometry.

- **Write a function that relates y to x.**
- **What do the variables x and y represent?**
- **Graph the function.**

a.

x	1	2	3	4	5
y	2π	4π	6π	8π	10π

b.

x	1	2	3	4	5
y	10	12	14	16	18

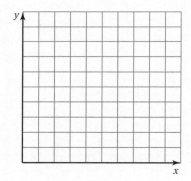

6.3 **Linear Functions** (continued)

c.

x	1	2	3	4	5
y	5	6	7	8	9

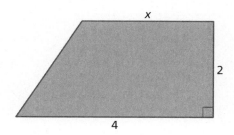

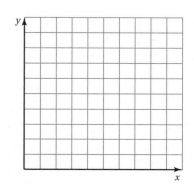

d.

x	1	2	3	4	5
y	28	40	52	64	76

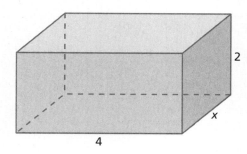

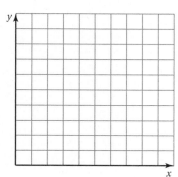

What Is Your Answer?

3. IN YOUR OWN WORDS How can you use a function to describe a linear pattern?

4. Describe the strategy you used to find the functions in Activities 1 and 2.

6.3 Practice
For use after Lesson 6.3

Use the graph or the table to write a linear function that relates _y_ to _x_.

1.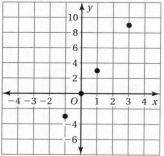

2.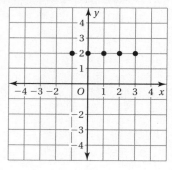

3.

x	0	1	2	3
y	5	7	9	11

4.

x	−2	0	2	4
y	−1	−2	−3	−4

5. The table shows the distance traveled _y_ (in miles) after _x_ hours.

x	0	2	4	6
y	0	120	240	360

a. Write a linear function that relates _y_ to _x_.

b. Graph the linear function.

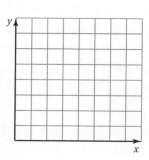

c. What is the distance traveled after three hours?

6.4 Comparing Linear and Nonlinear Functions
For use with Activity 6.4

Essential Question How can you recognize when a pattern in real life is linear or nonlinear?

1 ACTIVITY: Finding Patterns for Similar Figures

Work with a partner. Complete each table for the sequence of similar rectangles. Graph the data in each table. Decide whether each pattern is linear or nonlinear.

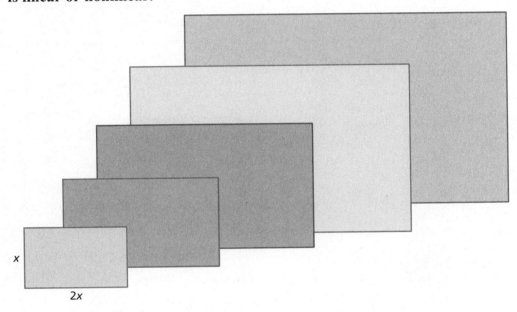

a. Perimeters of similar rectangles

x	1	2	3	4	5
P					

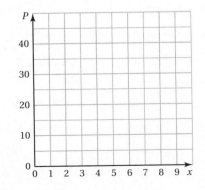

b. Areas of similar rectangles

x	1	2	3	4	5
A					

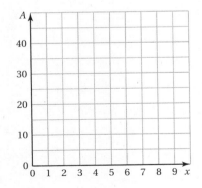

6.4 **Comparing Linear and Nonlinear Functions** (continued)

2 **ACTIVITY:** Comparing Linear and Nonlinear Functions

Work with a partner. Each table shows the height *h* (in feet) of a falling object at *t* seconds.

- Graph the data in each table.
- Decide whether each graph is linear or nonlinear.
- Compare the two falling objects. Which one has an increasing speed?

a. Falling parachute jumper

t	0	1	2	3	4
h	300	285	270	255	240

b. Falling bowling ball

t	0	1	2	3	4
h	300	284	236	156	44

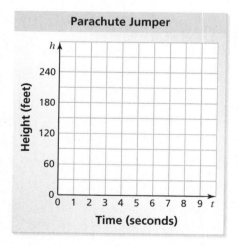

Parachute Jumper

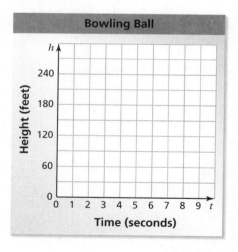

Bowling Ball

6.4 **Comparing Linear and Nonlinear Functions** (continued)

What Is Your Answer?

3. **IN YOUR OWN WORDS** How can you recognize when a pattern in real life is linear or nonlinear? Describe two real-life patterns: one that is linear and one that is nonlinear. Use patterns that are different from those described in Activities 1 and 2.

6.4 Practice

For use after Lesson 6.4

Graph the data in the table. Decide whether the graph is *linear* or *nonlinear*.

1.

x	−2	0	2	4
y	4	0	4	16

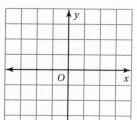

2.

x	−1	0	1	2
y	−1	1	3	5

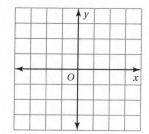

Does the graph represent a *linear* or nonlinear *function*? Explain.

3.

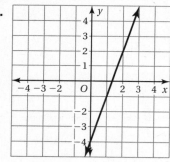

4.

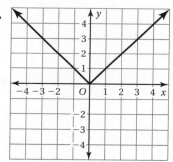

5. The table shows the area of a square with side length *x* inches. Does the table represent a linear or nonlinear function? Explain.

Side Length, x	1	2	3	4
Area, A	1	4	9	16

6.5 Analyzing and Sketching Graphs
For use with Activity 6.5

Essential Question How can you use a graph to represent relationships between quantities without using numbers?

 ACTIVITY: Interpreting a Graph

Work with a partner. Use the graph shown.

a. How is this graph different from the other graphs you have studied?

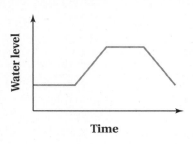

b. Write a short paragraph that describes how the water level changes over time.

c. What situation can this graph represent?

2 **ACTIVITY:** Matching Situations to Graphs

Work with a partner. You are riding your bike. Match each situation with the appropriate graph. Explain your reasoning.

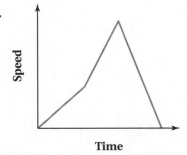

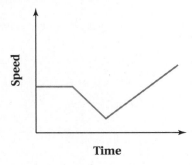

6.5 **Analyzing and Sketching Graphs** (continued)

C.

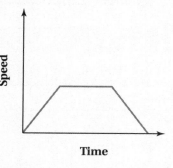

D.

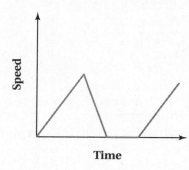

a. You gradually increase your speed, then ride at a constant speed along a bike path. You then slow down until you reach your friend's house.

b. You gradually increase your speed, then go down a hill. You then quickly come to a stop at an intersection.

c. You gradually increase your speed, then stop at a convenience store for a couple of minutes. You then continue to ride, gradually increasing your speed.

d. You ride at a constant speed, then go up a hill. Once on top of the hill, you gradually increase your speed.

3 **ACTIVITY: Comparing Graphs**

Work with a partner. The graphs represent the heights of a rocket and a weather balloon after they are launched.

a. How are the graphs similar? How are they different? Explain.

Graph A

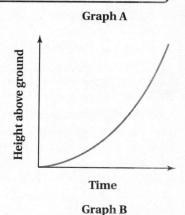

b. Compare the steepness of each graph.

Graph B

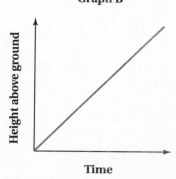

c. Which graph do you think represents the height of the rocket? Explain.

6.5 Analyzing and Sketching Graphs (continued)

4 ACTIVITY: Comparing Graphs

Work with a partner. The graphs represent the speeds of two cars. One car is approaching a stop sign. The other car is approaching a yield sign.

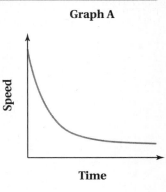

Graph A

  a. How are the graphs similar? How are they different? Explain.

  b. Compare the steepness of each graph.

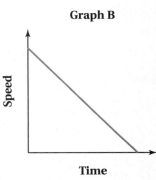

Graph B

  c. Which graph do you think represents the car approaching a stop sign? Explain.

What Is Your Answer?

  5. IN YOUR OWN WORDS How can you use a graph to represent relationships between quantities without using numbers?

  6. Describe a possible situation represented by the graph shown.

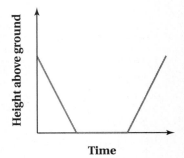

  7. Sketch a graph similar to the graphs in Activities 1 and 2. Exchange graphs with a classmate and describe a possible situation represented by the graph. Discuss the results.

Name _____ Date _____

Describe the relationship between the two quantities.

1. Sales

2. Bicycle

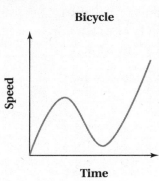

Sketch a graph that represents the situation.

3. You are texting a friend at a constant rate. You send the message then wait for a response. Once you receive a response, you begin texting a reply at a constant rate.

4. You cut your fingernails, let the nails grow back, and then cut them again.

Chapter 7 Fair Game Review

Complete the number sentence with <, >, or =.

1. 3.4 _____ 3.45

2. −6.01 _____ −6.1

3. 3.50 _____ 3.5

4. −0.84 _____ −0.91

Find three decimals that make the number sentence true.

5. −5.2 ≥ _____

6. 2.65 > _____

7. −3.18 ≤ _____

8. 0.03 < _____

9. The table shows the times of a 100-meter dash. Order the runners from first place to fifth place.

Runner	Time (seconds)
A	12.60
B	12.55
C	12.49
D	12.63
E	12.495

Chapter 7 **Fair Game Review** (continued)

Evaluate the expression.

10. $10^2 - 48 \div 6 + 25 \bullet 3$

11. $8\left(\dfrac{16}{4}\right) + 2^2 - 11 \bullet 3$

12. $\left(\dfrac{6}{3} + 4\right)^2 \div 4 \bullet 7$

13. $5(9 - 4)^2 - 3^2$

14. $5^2 - 2^2 \bullet 4^2 - 12$

15. $\left(\dfrac{50}{5^2}\right)^2 \div 4$

16. The table shows the numbers of students in 4 classes. The teachers are combining the classes and dividing the students in half to form two groups for a project. Write an expression to represent this situation. How many students are in each group?

Class	Students
1	24
2	32
3	30
4	28

Name_____ Date_____

7.1 Finding Square Roots
For use with Activity 7.1

Essential Question How can you find the dimensions of a square or a circle when you are given its area?

When you multiply a number by itself, you square the number.

| Symbol for squaring is the exponent 2. |

$4^2 = 4 \cdot 4$

$= 16$

4 squared is 16.

To "undo" this, take the *square root* of the number.

| Symbol for square root is a *radical sign*, $\sqrt{}$. |

$\sqrt{16} = \sqrt{4^2} = 4$

The square root of 16 is 4.

1 ACTIVITY: Finding Square Roots

Work with a partner. Use a square root symbol to write the side length of the square. Then find the square root. Check your answer by multiplying.

a. Sample: $s = \sqrt{121} =$ _____   **Check:**

Area = 121 ft^2

s

s

The length of each side of the square is _____.

b. Area = 81 yd^2

s

s

c. Area = 324 cm^2

s

s

d. Area = 361 mi^2

s

s

7.1 **Finding Square Roots** (continued)

e. Area = 225 mi²

s

s

f. Area = 2.89 in.²

s

s

g. Area = $\frac{4}{9}$ ft²

s

s

2 **ACTIVITY:** Using Square Roots

Work with a partner. Find the radius of each circle.

a.

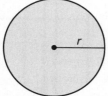

r

Area = 36π in.²

b.

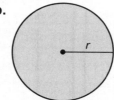

r

Area = π yd²

c.

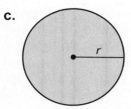

r

Area = 0.25π ft²

d.

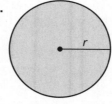

r

Area = $\frac{9}{16}$ π m²

3 **ACTIVITY:** The Period of a Pendulum

Work with a partner.

The period of a pendulum is the time (in seconds) it takes the pendulum to swing back *and* forth.

The period *T* is represented by $T = 1.1\sqrt{L}$, where *L* is the length of the pendulum (in feet).

Complete the table. Then graph the function on the next page. Is the function linear?

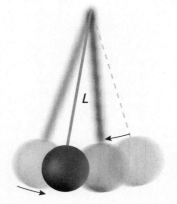

L

7.1 Finding Square Roots (continued)

L	1.00	1.96	3.24	4.00	4.84	6.25	7.29	7.84	9.00
T									

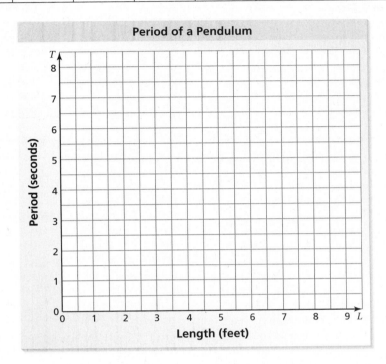

Period of a Pendulum

What Is Your Answer?

4. **IN YOUR OWN WORDS** How can you find the dimensions of a square or circle when you are given its area? Give an example of each. How can you check your answers?

Name _____ Date _____

7.1 **Practice**
For use after Lesson 7.1

Find the two square roots of the number.

1. 16

2. 100

3. 196

Find the square root(s).

4. $\sqrt{169}$

5. $\sqrt{\dfrac{4}{225}}$

6. $-\sqrt{12.25}$

Evaluate the expression.

7. $2\sqrt{36} + 9$

8. $8 - 11\sqrt{\dfrac{25}{121}}$

9. $3\left(\sqrt{\dfrac{125}{5}} - 8\right)$

10. A trampoline has an area of 49π square feet. What is the diameter of the trampoline?

7.2 Finding Cube Roots
For use with Activity 7.2

Essential Question How is the cube root of a number different from the square root of a number?

When you multiply a number by itself twice, you cube the number.

> Symbol for cubing is the exponent 3. → $4^3 = 4 \cdot 4 \cdot 4$
> $= 64$

4 cubed is 64.

To "undo" this, take the *cube root* of the number.

> Symbol for cube root is $\sqrt[3]{\ }$. → $\sqrt[3]{64} = \sqrt[3]{4^3} = 4$

The cube root of 64 is 4.

---

1 **ACTIVITY:** Finding Cube Roots

Work with a partner. Use a cube root symbol to write the edge length of the cube. Then find the cube root. Check your answer by multiplying.

a. **Sample:** $s = \sqrt[3]{343} = \sqrt[3]{7^3} = 7$ inches

Volume = 343 in.³

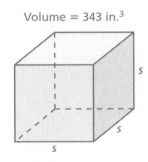

> **Check**
> $7 \cdot 7 \cdot 7 = 49 \cdot 7$
> $\qquad\qquad = 343 \checkmark$

The edge length of the cube is 7 inches.

b. Volume = 27 ft³

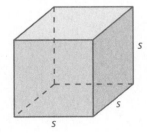

c. Volume = 125 m³

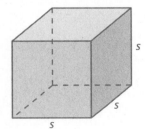

7.2 **Finding Cube Roots** (continued)

d. Volume = 0.001 cm³

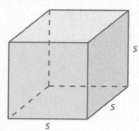

e. Volume = $\frac{1}{8}$ yd³

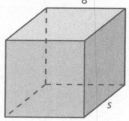

2 **ACTIVITY:** Use Prime Factorizations to Find Cube Roots

Work with a partner. Write the prime factorization of each number. Then use the prime factorization to find the cube root of the number.

a. 216

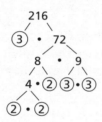

$216 = 3 \bullet 2 \bullet 3 \bullet 3 \bullet 2 \bullet 2$ Prime factorization

$= \left(3 \bullet \boxed{}\right) \bullet \left(3 \bullet \boxed{}\right) \bullet \left(3 \bullet \boxed{}\right)$ Commutative Property of Multiplication

$= \boxed{} \bullet \boxed{} \bullet \boxed{}$ Simplify.

The cube root of 216 is _____.

b. 1000

c. 3375

7.2 **Finding Cube Roots** (continued)

    d. STRUCTURE Does this procedure work for every number? Explain why or why not.

What Is Your Answer?

3. Complete each statement using *positive* or *negative*.

    a. A positive number times a positive number is a _____ number.

    b. A negative number times a negative number is a _____ number.

    c. A positive number multiplied by itself twice is a _____ number.

    d. A negative number multiplied by itself twice is a _____ number.

4. REASONING Can a negative number have a cube root? Give an example to support your explanation.

5. IN YOUR OWN WORDS How is the cube root of a number different from the square root of a number?

6. Give an example of a number whose square root and cube root are equal.

7. A cube has a volume of 13,824 cubic meters. Use a calculator to find the edge length.

7.2 Practice
For use after Lesson 7.2

Find the cube root.

1. $\sqrt[3]{27}$

2. $\sqrt[3]{8}$

3. $\sqrt[3]{-64}$

4. $\sqrt[3]{-\dfrac{125}{216}}$

Evaluate the expression.

5. $10 - \left(\sqrt[3]{12}\right)^3$

6. $2\sqrt[3]{512} + 10$

7. The volume of a cube is 1000 cubic inches. What is the edge length of the cube?

7.3 The Pythagorean Theorem
For use with Activity 7.3

Essential Question How are the lengths of the sides of a right triangle related?

Pythagoras was a Greek mathematician and philosopher who discovered one of the most famous rules in mathematics. In mathematics, a rule is called a **theorem**. So, the rule that Pythagoras discovered is called the Pythagorean Theorem.

Pythagoras
(c. 570–c. 490 B.C.)

1   ACTIVITY: Discovering the Pythagorean Theorem

Work with a partner.

a. On grid paper, draw any right triangle. Label the lengths of the two shorter sides a and b.

b. Label the length of the longest side c.

c. Draw squares along each of the three sides. Label the areas of the three squares a^2, b^2, and c^2.

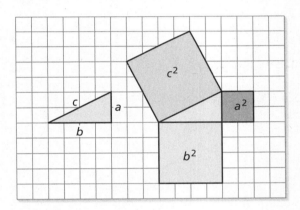

d. Cut out the three squares. Make eight copies of the right triangle and cut them out. Arrange the figures to form two identical larger squares.

e. MODELING The Pythagorean Theorem describes the relationship among a^2, b^2, and c^2. Use your result from part (d) to write an equation that describes this relationship.

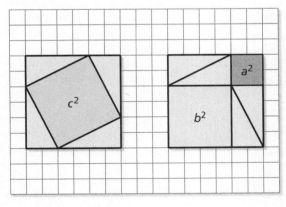

7.3 The Pythagorean Theorem (continued)

2 **ACTIVITY:** Using the Pythagorean Theorem in Two Dimensions

Work with a partner. Use a ruler to measure the longest side of each right triangle. Verify the result of Activity 1 for each right triangle.

a.
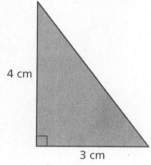
4 cm

3 cm

b.
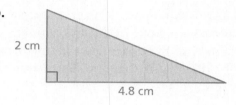
2 cm

4.8 cm

c.
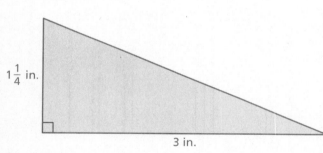
$1\frac{1}{4}$ in.

3 in.

d.
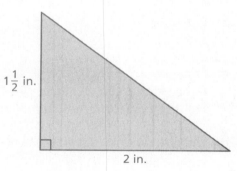
$1\frac{1}{2}$ in.

2 in.

7.3 The Pythagorean Theorem (continued)

3 ACTIVITY: Using the Pythagorean Theorem in Three Dimensions

Work with a partner. A guy wire attached 24 feet above ground level on a telephone pole provides support for the pole.

guy wire

a. **PROBLEM SOLVING** Describe a procedure that you could use to find the length of the guy wire without directly measuring the wire.

b. Find the length of the wire when it meets the ground 10 feet from the base of the pole.

What Is Your Answer?

4. **IN YOUR OWN WORDS** How are the lengths of the sides of a right triangle related? Give an example using whole numbers.

Name _____ Date _____

7.3 Practice
For use after Lesson 7.3

Find the missing length of the triangle.

1.

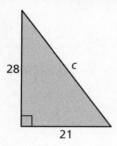

28   c

21

2.

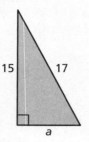

15   17

a

3.

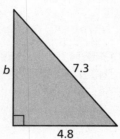

b   7.3

4.8

Find the missing length of the figure.

4.

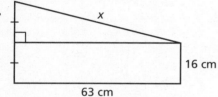

x

16 cm

63 cm

5.

x

13 m

35 m   5 m

6. In wood shop, you make a bookend that is in the shape
of a right triangle. What is the base *b* of the bookend?

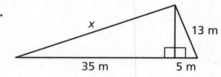

8 in.   10 in.

b

7.4 Approximating Square Roots
For use with Activity 7.4

Essential Question How can you find decimal approximations of square roots that are not rational?

1 ACTIVITY: Approximating Square Roots

Work with a partner. Archimedes was a Greek mathematician, physicist, engineer, inventor, and astronomer. He tried to find a rational number whose square is 3. Two that he tried were $\dfrac{265}{153}$ and $\dfrac{1351}{780}$.

    a. Are either of these numbers equal to $\sqrt{3}$? Explain.

    b. Use a calculator to approximate $\sqrt{3}$. Write the number on a piece of paper. Enter it into the calculator and square it. Then subtract 3. Do you get 0? What does this mean?

    c. The value of $\sqrt{3}$ is between which two integers?

    d. Tell whether the value of $\sqrt{3}$ is between the given numbers. Explain your reasoning.

| 1.7 and 1.8 | 1.72 and 1.73 | 1.731 and 1.732 |

2 ACTIVITY: Approximating Square Roots Geometrically

Work with a partner. Refer to the square on the number line below.

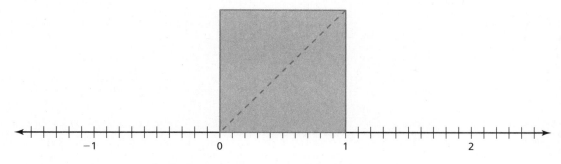

    a. What is the length of the diagonal of the square?

    b. Copy the square and its diagonal onto a piece of transparent paper. Rotate it about zero on the number line so that the diagonal aligns with the number line. Use the number line to estimate the length of the diagonal.

7.4   **Approximating Square Roots** (continued)

    c. STRUCTURE How do you think your answers in parts (a) and (b) are related?

3   **ACTIVITY:** Approximating Square Roots Geometrically

Work with a partner.

    a. Use grid paper and the given scale to draw a horizontal line segment 1 unit in length. Draw your segment near the bottom of the grid. Label this segment *AC*.

    b. Draw a vertical line segment 2 units in length. Draw your segment near the left edge of the grid. Label this segment *DC*.

    c. Set the point of a compass on *A*.   **d.** Use the Pythagorean Theorem to
        Set the compass to 2 units. Swing   find the length of segment *BC*.
        the compass to intersect segment
        DC. Label this intersection as *B*.

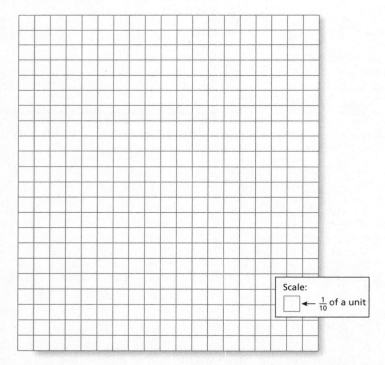

Scale:
⬜ ← $\frac{1}{10}$ of a unit

    e. Use the grid paper to approximate $\sqrt{3}$ to the nearest tenth.

7.4 **Approximating Square Roots** (continued)

4. Compare your approximation in Activity 3 with your results from Activity 1.

What Is Your Answer?

5. Repeat Activity 3 for a triangle in which segment *AC* is 2 units and segment *BA* is 3 units. Use the Pythagorean Theorem to find the length of segment *BC*. Use the grid paper to approximate $\sqrt{5}$ to the nearest tenth.

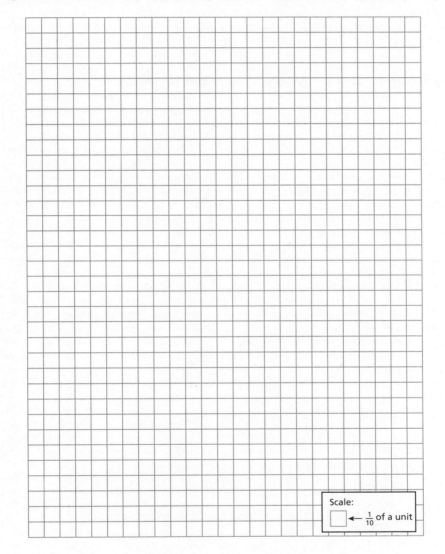

Scale:

□ ← $\frac{1}{10}$ of a unit

6. **IN YOUR OWN WORDS** How can you find decimal approximations of square roots that are not rational?

7.4 Practice
For use after Lesson 7.4

Classify the real number.

1. $\sqrt{14}$

2. $-\dfrac{3}{7}$

3. $\dfrac{153}{3}$

Estimate the square root to the nearest (a) integer and (b) tenth.

4. $\sqrt{8}$

5. $\sqrt{60}$

6. $-\sqrt{\dfrac{172}{25}}$

Which number is greater? Explain.

7. $\sqrt{88}$, 12

8. $-\sqrt{18}$, -6

9. 14.5, $\sqrt{220}$

10. The velocity in meters per second of a ball that is dropped from a window at a height of 10.5 meters is represented by the equation $v = \sqrt{2(9.8)(10.5)}$. Estimate the velocity of the ball. Round your answer to the nearest tenth.

Name_____ Date_____

Practice
For use after Extension 7.4

Write the decimal as a fraction or a mixed number.

1. $0.\overline{3}$

2. $-0.\overline{2}$

3. $1.\overline{7}$

4. $-2.\overline{6}$

5. $0.4\overline{6}$

6. $-1.8\overline{3}$

Name _____ Date _____

Extension 7.4 **Practice** (continued)

7. $-0.7\overline{3}$

8. $0.\overline{18}$

9. $-3.\overline{24}$

10. $1.0\overline{9}$

11. The length of a pencil is $1.5\overline{6}$ inches. Represent the length of the pencil as a mixed number.

7.5 Using the Pythagorean Theorem
For use with Activity 7.5

Essential Question In what other ways can you use the Pythagorean Theorem?

The *converse* of a statement switches the hypothesis and the conclusion.

Statement:	Converse of the statement:
If p, then q.	If q, then p.

1 ACTIVITY: Analyzing Converses of Statements

Work with a partner. Write the converse of the true statement. Determine whether the converse is *true* or *false*. If it is true, justify your reasoning. If it is false, give a counterexample.

  a. If $a = b$, then $a^2 = b^2$.

   Converse: _____

  b. If $a = b$, then $a^3 = b^3$.

   Converse: _____

  c. If one figure is a translation of another figure, then the figures are congruent.

   Converse: _____

  d. If two triangles are similar, then the triangles have the same angle measures.

   Converse: _____

Is the converse of a true statement always true? always false? Explain.

7.5 **Using the Pythagorean Theorem** (continued)

2 **ACTIVITY:** The Converse of the Pythagorean Theorem

Work with a partner. The converse of the Pythagorean Theorem states: "If the equation $a^2 + b^2 = c^2$ is true for the side lengths of a triangle, then the triangle is a right triangle."

    a. Do you think the converse of the Pythagorean Theorem is *true* or *false*? How could you use deductive reasoning to support your answer?

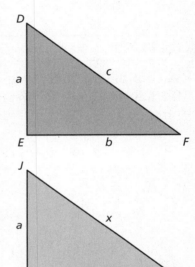

    b. Consider $\triangle DEF$ with side lengths a, b, and c, such that $a^2 + b^2 = c^2$. Also consider $\triangle JKL$ with leg lengths a and b, where $\angle K = 90°$.

       • What does the Pythagorean Theorem tell you about $\triangle JKL$?

       • What does this tell you about c and x?

       • What does this tell you about $\triangle DEF$ and $\triangle JKL$?

       • What does this tell you about $\angle E$?

       • What can you conclude?

7.5 **Using the Pythagorean Theorem** (continued)

3 **ACTIVITY:** Developing the Distance Formula

Work with a partner. Follow the steps below to write a formula that you can use to find the distance between and two points in a coordinate plane.

Step 1: Choose two points in the coordinate plane that do not lie on the same horizontal or vertical line. Label the points (x_1, y_1) and (x_2, y_2).

Step 2: Draw a line segment connecting the points. This will be the hypotenuse of a right triangle.

Step 3: Draw horizontal and vertical line segments from the points to form the legs of the right triangle.

Step 4: Use the *x*-coordinates to write an expression for the length of the horizontal leg.

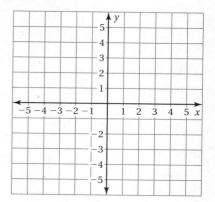

Step 5: Use the *y*-coordinates to write an expression for the length of the vertical leg.

Step 6: Substitute the expressions for the lengths of the legs into the Pythagorean Theorem.

Step 7: Solve the equation in Step 6 for the hypotenuse *c*.

What does the length of the hypotenuse tell you about the two points?

What Is Your Answer?

4. IN YOUR OWN WORDS In what other ways can you use the Pythagorean Theorem?

5. What kind of real-life problems do you think the converse of the Pythagorean Theorem can help you solve?

Name _____ Date _____

Tell whether the triangle with the given side lengths is a right triangle.

1.

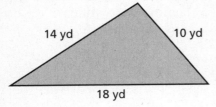

14 yd
10 yd
18 yd

2.

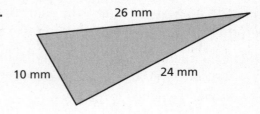

26 mm
10 mm
24 mm

3. 4 m, 4.2 m, 5.8 m

4. 31 in., 35 in., 16 in.

Find the distance between the two points.

5. $(2, 1), (-3, 6)$

6. $(-6, -4), (2, 2)$

7. $(1, -7), (4, -5)$

8. $(-9, 3), (-5, -8)$

9. The cross-section of a wheelchair ramp is shown. Does the ramp form a right triangle?

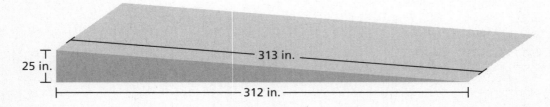

25 in.
313 in.
312 in.

Name_____ Date_____

Find the area of the figure.

1.

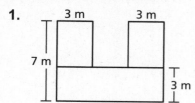

3 m   3 m

7 m

3 m

9 m

2.

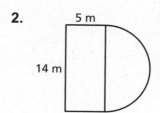

5 m

14 m

3.

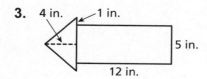

4 in.   1 in.

5 in.

12 in.

4.

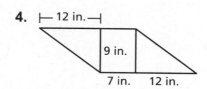

⊢ 12 in. ⊣

9 in.

7 in.   12 in.

5.

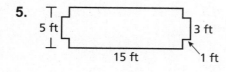

5 ft

3 ft

15 ft   1 ft

6.

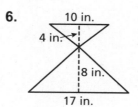

10 in.

4 in.

8 in.

17 in.

7. You are carpeting 2 rooms of your house.
The carpet costs $1.48 per square foot.
How much does it cost to carpet the rooms?

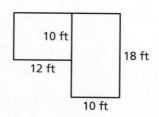

10 ft

12 ft

18 ft

10 ft

Chapter 8 **Fair Game Review** (continued)

Find the area of the circle.

8.

20 in.

9.

6 m

10.

12 cm

11.

14 ft

12.

25 yd

13.
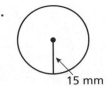
15 mm

14. Find the area of the shaded region.
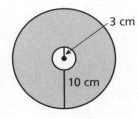
3 cm

10 cm

Name_____   Date_____

8.1 Volumes of Cylinders
For use with Activity 8.1

Essential Question   How can you find the volume of a cylinder?

---

1 ACTIVITY: Finding a Formula Experimentally

Work with a partner.

   a. Find the area of the face of a coin.

   b. Find the volume of a stack of a dozen coins.

   c. Write a formula for the volume of a cylinder.

 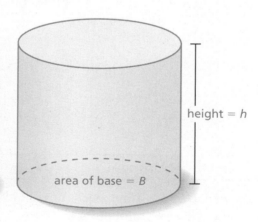

height = h

area of base = B

8.1 **Volumes of Cylinders** (continued)

2 ACTIVITY: Making a Business Plan

Work with a partner. You are planning to make and sell three different sizes of cylindrical candles. You buy 1 cubic foot of candle wax for $20 to make 8 candles of each size.

    a. Design the candles. What are the dimensions of each size of candle?

    b. You want to make a profit of $100. Decide on a price for each size of candle.

    c. Did you set the prices so that they are proportional to the volume of each size of candle? Why or why not?

3 ACTIVITY: Science Experiment

Work with a partner. Use the diagram to describe how you can find the volume of a small object.

8.2 Volumes of Cones
For use with Activity 8.2

Essential Question How can you find the volume of a cone?

You already know how the volume of a pyramid relates to the volume of a prism. In this activity, you will discover how the volume of a cone relates to the volume of a cylinder.

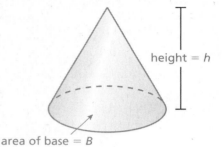

height = h

area of base = B

---

1 ACTIVITY: Finding a Formula Experimentally

Work with a partner. Use a paper cup that is shaped like a cone.

- Estimate the height of the cup.

- Trace the top of the cup on a piece of paper. Find the diameter of the circle.

- Use these measurements to draw a net for a cylinder with the same base and height as the paper cup.

- Cut out the net. Then fold and tape it to form an open cylinder.

- Fill the paper cup with rice. Then pour the rice into the cylinder. Repeat this until the cylinder is full. How many cones does it take to fill the cylinder?

- Use your result to write a formula for the volume of a cone.

8.2 **Volumes of Cones** (continued)

2 **ACTIVITY:** Summarizing Volume Formulas

Work with a partner. You can remember the volume formulas for prisms, cylinders, pyramids, and cones with just two concepts.

Volumes of Prisms and Cylinders

Volume = | Area of base | × | |

Volumes of Pyramids and Cones

Volume = | | | Volume of prism or cylinder with same base and height |

Make a list of all the formulas you need to remember to find the area of a base. Talk about strategies for remembering these formulas.

3 **ACTIVITY:** Volumes of Oblique Solids

Work with a partner. Think of a stack of paper. When you adjust the stack so that the sides are oblique (slanted), do you change the volume of the stack? If the volume of the stack does not change, then the formulas for volumes of right solids also apply to oblique solids.

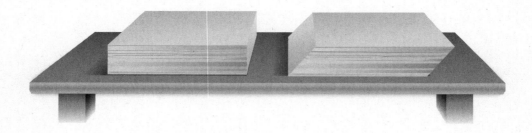

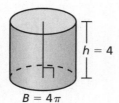
$h = 4$
$B = 4\pi$

Right Cylinder

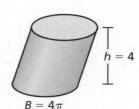
$h = 4$
$B = 4\pi$

Oblique Cylinder

8.2 **Volumes of Cones** (continued)

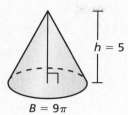

$h = 5$

$B = 9\pi$

Right Cone

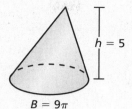

$h = 5$

$B = 9\pi$

Oblique Cone

What Is Your Answer?

4. **IN YOUR OWN WORDS** How can you find the volume of a cone?

5. Describe the intersection of the plane and the cone. Then explain how to find the volume of each section of the solid.

a.

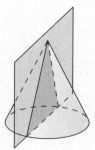

b.

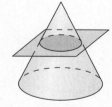

Name _____ Date _____

Find the volume of the cone. Round your answer to the nearest tenth.

1.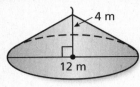

2.

3.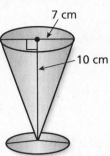

Find the missing dimension of the cone. Round your answer to the nearest tenth.

4. Volume = 300π mm^3

5. Volume = 78.5 cm^3

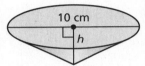

6. What is the volume of the catch and click cone?

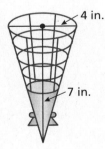

8.3 Volumes of Spheres
For use with Activity 8.3

Essential Question   How can you find the volume of a sphere?

A **sphere** is the set of all points in space that are the same distance from a point called the *center*. The *radius r* is the distance from the center to any point on the sphere.

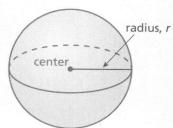

radius, *r*

center

A sphere is different from the other solids you have studied so far because it does not have a base. To discover the volume of a sphere, you can use an activity similar to the one in the previous section.

---

1   ACTIVITY: Exploring the Volume of a Sphere

Work with a partner. Use a plastic ball similar to the one shown.

• Estimate the diameter and the radius of the ball.

• Use these measurements to draw a net for a cylinder with a diameter and a height equal to the diameter of the ball. How is the height *h* of the cylinder related to the radius *r* of the ball? Explain.

• Cut out the net. Then fold and tape it to form an open cylinder. Make two marks on the cylinder that divide it into thirds, as shown.

• Cover the ball with aluminum foil or tape. Leave one hole open. Fill the ball with rice. Then pour the rice into the cylinder. What fraction of the cylinder is filled with rice?

8.3   **Volume of Spheres** (continued)

2   **ACTIVITY:** Deriving the Formula for the Volume of a Sphere

Work with a partner. Use the results from Activity 1 and the formula for the volume of a cylinder to complete the steps.

$V = \pi r^2 h$ Write formula for volume of a cylinder.

$= \dfrac{\boxed{}}{\boxed{}} \pi r^2 h$ Multiply by $\dfrac{\boxed{}}{\boxed{}}$ because the volume of a sphere is $\dfrac{\boxed{}}{\boxed{}}$ of the volume of the cylinder.

$= \dfrac{\boxed{}}{\boxed{}} \pi r^2 \boxed{}$ Substitute $\boxed{}$ for h.

$= \dfrac{\boxed{}}{\boxed{}} \pi \boxed{}$ Simplify.

3   **ACTIVITY:** Deriving the Formula for the Volume of a Sphere

Work with a partner. Imagine filling the inside of a sphere with n small pyramids. The vertex of each pyramid is at the center of the sphere and the height of each pyramid is approximately equal to r, as shown. Complete the steps. $\left(\text{The surface area of a sphere is equal to } 4\pi r^2.\right)$

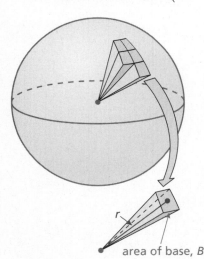

area of base, B

$V = \dfrac{1}{3}Bh$ Write formula for volume of a pyramid.

$= n\dfrac{1}{3}B\boxed{}$ Multiply by the number of small pyramids n and substitute $\boxed{}$ for h.

$= \dfrac{1}{3}\left(4\pi r^2\right)\boxed{}$ $4\pi r^2 \approx n \bullet \boxed{}$.

Show how this result is equal to the result in Activity 2.

8.3 **Volume of Spheres** (continued)

What Is Your Answer?

4. **IN YOUR OWN WORDS** How can you find the volume of a sphere?

5. Describe the intersection of the plane and the sphere. Then explain how to find the volume of each section of the solid.

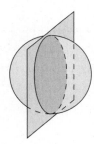

8.3 **Practice**
For use after Lesson 8.3

Find the volume of the sphere. Round your answer to the nearest tenth.

1.

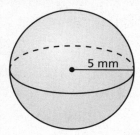

5 mm

2.

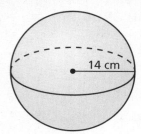

14 cm

3.

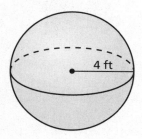

4 ft

4.

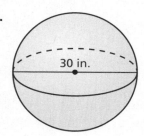

30 in.

5. Find the volume of the exercise ball. Round your answer to the nearest tenth.

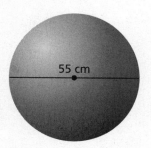

55 cm

Name_____ Date _____

8.4 Surface Areas and Volumes of Similar Solids
For use with Activity 8.4

Essential Question When the dimensions of a solid increase by a factor of k, how does the surface area change? How does the volume change?

1 **ACTIVITY:** Comparing Surface Areas and Volumes

Work with a partner. Complete the table. Describe the pattern. Are the dimensions proportional? Explain your reasoning.

a.

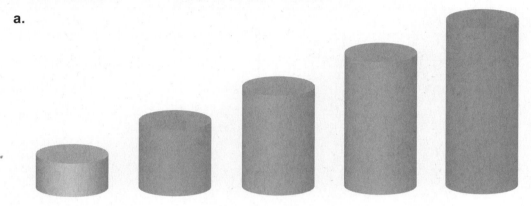

Radius	1	1	1	1	1
Height	1	2	3	4	5
Surface Area					
Volume					

8.4 Surface Areas and Volumes of Similar Solids (continued)

b.

Radius	1	2	3	4	5
Height	1	2	3	4	5
Surface Area					
Volume					

2 ACTIVITY: Comparing Surface Areas and Volumes

Work with a partner. Complete the table. Describe the pattern. Are the dimensions proportional? Explain.

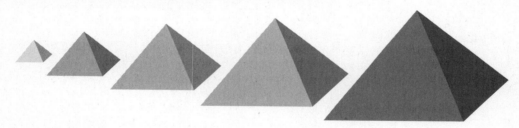

Base Side	6	12	18	24	30
Height	4	8	12	16	20
Slant Height	5	10	15	20	25
Surface Area					
Volume					

8.4 **Surface Areas and Volumes of Similar Solids** (continued)

What Is Your Answer?

3. **IN YOUR OWN WORDS** When the dimensions of a solid increase by a factor of k, how does the surface area change?

4. **IN YOUR OWN WORDS** When the dimensions of a solid increase by a factor of k, how does the volume change?

5. **REPEATED REASONING** All the dimensions of a prism increase by a factor of 5.

   a. How many times greater is the surface area? Explain.

   | 5 | 10 | 25 | 125 |

   b. How many times greater is the volume? Explain.

   | 5 | 10 | 25 | 125 |

8.4 Practice
For use after Lesson 8.4

Determine whether the solids are similar.

1.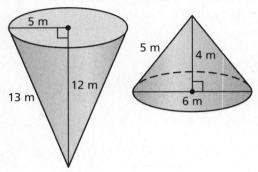

2.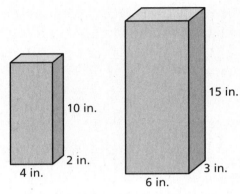

The solids are similar. Find the missing dimension(s).

3.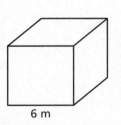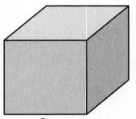

4.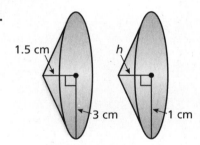

The solids are similar. Find the surface area S or volume V of the shaded solid.

5.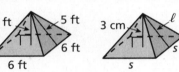

6 m   8 m

Surface Area = 198 m²

6.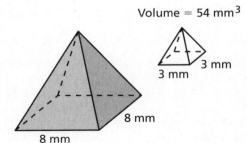

Volume = 54 mm³

3 mm

3 mm

8 mm

8 mm

Name_____ Date_____

Chapter 9   Fair Game Review

Plot the ordered pair in a coordinate plane. Describe the location of the point.

1. $(2, 1)$

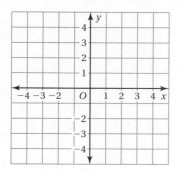

2. $(-3, 3)$

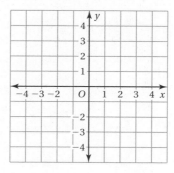

3. $(4, -2)$

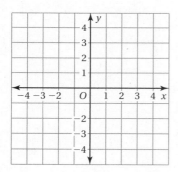

4. $(-1, -1)$

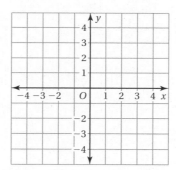

5. Describe the location of the vertices of the triangle.

Chapter 9 **Fair Game Review** (continued)

Write in slope-intercept form an equation of the line that passes through the given points.

6. $(-2, -2), (1, 7)$

7. $(5, -1), (-5, 11)$

8. $(-20, -8), (5, 12)$

9. $(6, -11), (-3, 1)$

10. $(-1, -3), (2, 6)$

11. $(-3, 6), (4, -8)$

Name_____   Date_____

9.1 Scatter Plots
For use with Activity 9.1

Essential Question How can you construct and interpret a scatter plot?

1 ACTIVITY: Constructing a Scatter Plot

Work with a partner. The weights x (in ounces) and circumferences C (in inches) of several sports balls are shown.

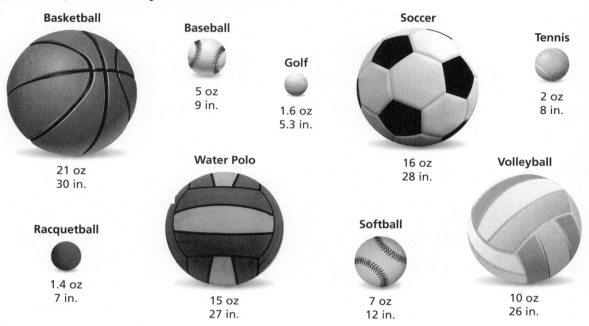

Basketball
21 oz
30 in.

Baseball
5 oz
9 in.

Golf
1.6 oz
5.3 in.

Soccer
16 oz
28 in.

Tennis
2 oz
8 in.

Water Polo
15 oz
27 in.

Volleyball
10 oz
26 in.

Racquetball
1.4 oz
7 in.

Softball
7 oz
12 in.

a. Choose a scale for the horizontal axis and the vertical axis of the coordinate plane shown.

b. Write the weight x and circumference C of each ball as an ordered pair. Then plot the ordered pairs in the coordinate plane.

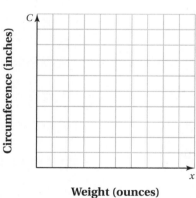

c. Describe the relationship between weight and circumference. Are any of the points close together?

Name _____ Date _____

9.1 **Scatter Plots** (continued)

d. In general, do you think you can describe this relationship as *positive* or *negative*? *linear* or *nonlinear*? Explain.

e. A bowling ball has a weight of 225 ounces and a circumference of 27 inches. Describe the location of the ordered pair that represents this data point in the coordinate plane. How does this point compare to the others? Explain your reasoning.

2 **ACTIVITY: Constructing a Scatter Plot**

Work with a partner. The table shows the number of absences and the final grade for each student in a sample.

a. Write the ordered pairs from the table. Then plot them in the coordinate plane.

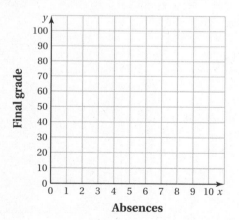

Absences	Final Grade
0	95
3	88
2	90
5	83
7	79
9	70
4	85
1	94
10	65
8	75

b. Describe the relationship between absences and final grade. How is this relationship similar to the relationship between weight and circumference in Activity 1? How is it different?

9.1 **Scatter Plots** (continued)

   c. **MODELING** A student has been absent 6 days. Use the data to predict the student's final grade. Explain how you found your answer.

3 **ACTIVITY:** Identifying Scatter Plots

Work with a partner. Match the data sets with the most appropriate scatter plot. Explain your reasoning.

   a. month of birth and birth weight for infants at a day care

   b. quiz score and test score of each student in a class

   c. age and value of laptop computers

i.

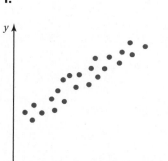

ii.

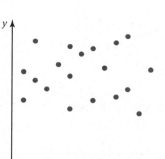

iii.

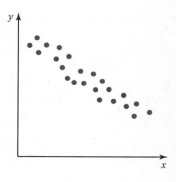

What Is Your Answer?

   4. How would you define the term *scatter plot*?

   5. IN YOUR OWN WORDS How can you construct and interpret a scatter plot?

9.1 Practice
For use after Lesson 9.1

1. The scatter plot shows the participation in a bowling league over eight years.

   a. About how many people were in the league in 2008?

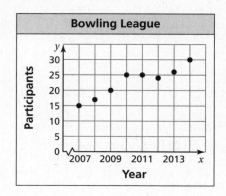

   b. Describe the relationship shown by the data.

Describe the relationship between the data. Identify any outliers, gaps, or clusters.

2.

3.

4.

Name_____ Date_____

9.2 Lines of Fit
For use with Activity 9.2

Essential Question How can you use data to predict an event?

---

1 ACTIVITY: Representing Data by a Linear Equation

Work with a partner. You have been working on a science project for 8 months. Each month, you measured the length of a baby alligator.

The table shows your measurements.

September ↓   April ↓

Month, x	0	1	2	3	4	5	6	7
Length (in.), y	22.0	22.5	23.5	25.0	26.0	27.5	28.5	29.5

Use the following steps to predict the baby alligator's length next September.

a. Graph the data in the table.

b. Draw a line that you think best approximates the points.

c. Write an equation for your line.

d. **MODELING** Use the equation to predict the baby alligator's length next September.

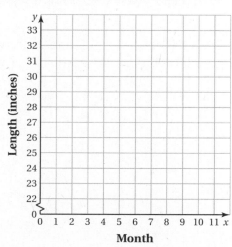

9.2 **Lines of Fit** (continued)

2 **ACTIVITY:** Representing Data by a Linear Equation

Work with a partner. You are a biologist and study bat populations.

You are asked to predict the number of bats that will be living in an abandoned mine in 3 years.

To start, you find the number of bats that have been living in the mine during the past 8 years.

The table shows the results of your research.

7 years ago

this year

Year, x	0	1	2	3	4	5	6	7
Bats (thousands), y	327	306	299	270	254	232	215	197

Use the following steps to predict the number of bats that will be living in the mine after 3 years.

   a. Graph the data in the table.

   b. Draw a line that you think best approximates the points.

   c. Write an equation for your line.

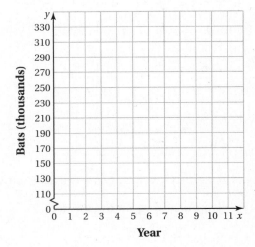

   d. MODELING Use the equation to predict the number of bats in 3 years.

9.2 **Lines of Fit** (continued)

What Is Your Answer?

3. **IN YOUR OWN WORDS** How can you use data to predict an event?

4. **MODELING** Use the Internet or some other reference to find data that appear to have a linear pattern. List the data in a table and graph the data. Use an equation that is based on the data to predict a future event.

9.2 Practice
For use after Lesson 9.2

1. The table shows the money you owe to pay off a credit card bill over five months.

   a. Make a scatter plot of the data and draw a line of fit.

   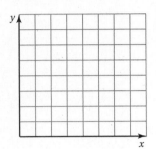

Months, x	Money owed (dollars), y
1	1200
2	1000
3	850
4	600
5	410

   b. Write an equation of the line of fit.

   c. Interpret the slope and *y*-intercept of the line of fit.

   d. Predict the amount of money you will owe in six months.

Use a graphing calculator to find an equation of the line of best fit. Identify and interpret the correlation coefficient.

2.

x	−8	−6	−4	−2	0	2	4	6	8
y	10	7	1	0	−3	−5	−4	−14	−11

3.

x	1	2	3	4	5	6	7	8
y	8	6	4	2	0	2	4	6

9.3 Two-Way Tables
For use with Activity 9.3

Essential Question How can you read and make a two-way table?

Two categories of data can be displayed in a *two-way table*.

1 ACTIVITY: Reading a Two-Way Table

Work with a partner. You are the manager of a sports shop. The two-way table shows the numbers of soccer T-shirts that your shop has left in stock at the end of the season.

		S	M	L	XL	XXL	Total
		T-Shirt Size					
Color	Blue/White	5	4	1	0	2	
	Blue/Gold	3	6	5	2	0	
	Red/White	4	2	4	1	3	
	Black/White	3	4	1	2	1	
	Black/Gold	5	2	3	0	2	
	Total						65

a. Complete the totals for the rows and columns.

b. Are there any black-and-gold XL T-shirts in stock? Justify your answer.

c. The numbers of T-shirts you ordered at the beginning of the season are shown below. Complete the two-way table.

		S	M	L	XL	XXL	Total
		T-Shirt Size					
Color	Blue/White	5	6	7	6	5	
	Blue/Gold	5	6	7	6	5	
	Red/White	5	6	7	6	5	
	Black/White	5	6	7	6	5	
	Black/Gold	5	6	7	6	5	
	Total						

9.3 **Two-Way Tables** (continued)

d. **REASONING** How would you alter the numbers of T-shirts you order for next season? Explain your reasoning.

2 **ACTIVITY:** Analyzing Data

Work with a partner. The three-dimensional two-way table shows information about the numbers of hours students at a high school work at part-time jobs during the school year.

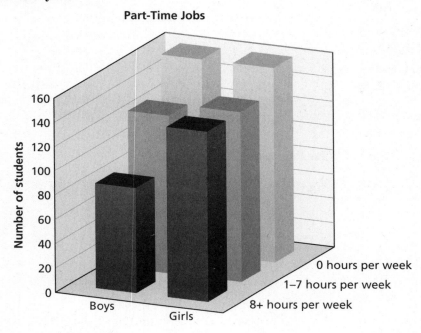

Part-Time Jobs

a. Make a two-way table showing the data. Use estimation to find the entries in your table.

9.3 Two-Way Tables (continued)

b. Write two observations you can make that summarize the data in your table.

c. REASONING A newspaper article claims that more boys than girls drop out of high school to work full-time. Do the data support this claim? Explain your reasoning.

What Is Your Answer?

3. IN YOUR OWN WORDS How can you read and make a two-way table?

4. Find a real-life data set that can be represented by a two-way table. Then make a two-way table for the data set.

9.3 Practice
For use after Lesson 9.3

1. You randomly survey students in a school about whether they got the flu after receiving a flu shot. The results of the survey are shown in the two-way table.

   a. How many of the students in the survey received a flu shot and still got the flu?

   b. Find and interpret the marginal frequencies for the survey.

		Flu Shot		
		Yes	No	Total
Flu	Yes	8	13	
	No	27	32	
	Total			

2. You randomly survey students in a school about whether they eat breakfast at home or at school.

   Grade 6 Students: 28 eat breakfast at home, 12 eat breakfast at school

   Grade 7 Students: 15 eat breakfast at home, 15 eat breakfast at school

   Grade 8 Students: 9 eat breakfast at home, 21 eat breakfast at school

   a. Make a two-way table that includes the marginal frequencies.

   b. For each grade level, what percent of the students in the survey eat breakfast at home? eat breakfast at school? Organize the results in a two-way table. Explain what one of the entries represents.

9.4 Choosing a Data Display
For use with Activity 9.4

Essential Question How can you display data in a way that helps you make decisions?

---

1 ACTIVITY: Displaying Data

Work with a partner. Analyze and display each data set in a way that best describes the data. Explain your choice of display.

a. **ROADKILL** A group of schools in New England participated in a 2-month study and reported 3962 dead animals.

Birds: 307
Mammals: 2746
Amphibians: 145
Reptiles: 75
Unknown: 689

b. **BLACK BEAR ROADKILL** The data below show the numbers of black bears killed on a state's roads from 1993 to 2012.

1993:	30	2003:	74
1994:	37	2004:	88
1995:	46	2005:	82
1996:	33	2006:	109
1997:	43	2007:	99
1998:	35	2008:	129
1999:	43	2009:	111
2000:	47	2010:	127
2001:	49	2011:	141
2002:	61	2012:	135

9.4 **Choosing a Data Display** (continued)

c. **RACCOON ROADKILL** A 1-week study along a 4-mile section of road found the following weights (in pounds) of raccoons that had been killed by vehicles.

13.4	14.8	17.0	12.9	21.3	21.5	16.8	14.8
15.2	18.7	18.6	17.2	18.5	9.4	19.4	15.7
14.5	9.5	25.4	21.5	17.3	19.1	11.0	12.4
20.4	13.6	17.5	18.5	21.5	14.0	13.9	19.0

d. What do you think can be done to minimize the number of animals killed by vehicles?

2 **ACTIVITY:** Statistics Project

ENDANGERED SPECIES PROJECT Use the Internet or some other reference to write a report about an animal species that is (or has been) endangered. Include graphical displays of the data you have gathered.

Sample: Florida Key Deer In 1939, Florida banned the hunting of Key deer. The numbers of Key deer fell to about 100 in the 1940s.

About half of Key deer deaths are due to vehicles.

Name_____ Date _____

9.4 **Choosing a Data Display** (continued)

In 1947, public sentiment was stirred by 11-year-old Glenn Allen from Miami. Allen organized Boy Scouts and others in a letter-writing campaign that led to the establishment of the National Key Deer Refuge in 1957. The approximately 8600-acre refuge includes 2280 acres of designated wilderness.

The Key Deer Refuge has increased the population of Key deer. A recent study estimated the total Key deer population to be approximately 800.

One of two Key deer wildlife underpasses on Big Pine Key.

What Is Your Answer?

3. **IN YOUR OWN WORDS** How can you display data in a way that helps you make decisions? Use the Internet or some other reference to find examples of the following types of data displays.

- Bar graph
- Circle graph
- Scatter plot
- Stem-and-leaf plot
- Box-and-whisker plot

Copyright © Big Ideas Learning, LLC
All rights reserved.

Big Ideas Math Blue **199**
Record and Practice Journal

9.4 Practice
For use after Lesson 9.4

Choose an appropriate data display for the situation. Explain your reasoning.

1. the number of people that donated blood over the last 5 years

2. percent of class participating in school clubs

Explain why the data display is misleading.

3.

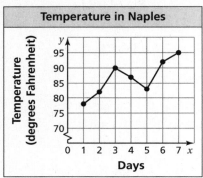

4.

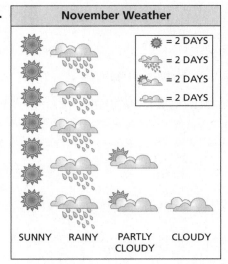

5. A team statistician wants to use a data display to show the points scored per game during the season. Choose an appropriate data display for the situation. Explain your reasoning.

Chapter 10 **Fair Game Review**

Evaluate the expression.

1. $2 + 1 \bullet 4^2 - 12 \div 3$

2. $8^2 \div 16 \bullet 2 - 5$

3. $7(9 - 3) + 6^2 \bullet 10 - 8$

4. $3 \bullet 5 - 10 + 9(2 + 1)^2$

5. $8(6 + 5) - (9^2 + 3) \div 7$

6. $5\big[3(12 - 8)\big] - 6 \bullet 8 + 2^2$

7. $4 + 4 + 5 \times 2 \times 5 + (3 + 3 + 3) \times 6 \times 6 + 2 + 2$

   a. Evaluate the expression.

   b. Rewrite the expression using what you know about order of operations.
   Then evaluate.

Chapter 10 Fair Game Review (continued)

Find the product or quotient.

8. $3.92 \cdot 0.6$

9. $0.78 \cdot 0.13$

10. $\begin{array}{r} 5.004 \\ \times\ \ 1.2 \\ \hline \end{array}$

11. $6.3 \div 0.7$

12. $2.25 \div 1.5$

13. $0.003\overline{)8.1}$

14. Grapes cost \$1.98 per pound. You buy 3.5 pounds of grapes. How much do you pay for the grapes?

Name_____ Date_____

Exponents
For use with Activity 10.1

Essential Question How can you use exponents to write numbers?

The expression 3^5 is called a *power*. The *base* is 3. The *exponent* is 5.

base → 3^5 ← exponent

1 ACTIVITY: Using Exponent Notation

Work with a partner.

a. Complete the table.

Power	Repeated Multiplication Form	Value
$(-3)^1$		
$(-3)^2$		
$(-3)^3$		
$(-3)^4$		
$(-3)^5$		
$(-3)^6$		
$(-3)^7$		

b. **REPEATED REASONING** Describe what is meant by the expression $(-3)^n$.

How can you find the value of $(-3)^n$?

10.1 Exponents (continued)

2 **ACTIVITY:** Using Exponent Notation

Work with a partner.

a. The cube at the right has $3 in each of its small cubes. Write a power that represents the total amount of money in the large cube.

b. Evaluate the power to find the total amount of money in the large cube.

3 **ACTIVITY:** Writing Powers as Whole Numbers

Work with a partner. Write each distance as a whole number. Which numbers do you know how to write in words? For instance, in words, 10^3 is equal to *one thousand*.

a. 10^{26} meters:
   diameter of observable universe

b. 10^{21} meters:
   diameter of Milky Way galaxy

c. 10^{16} meters:
   diameter of solar system

d. 10^7 meters:
   diameter of Earth

e. 10^6 meters:
   length of Lake Erie shoreline

f. 10^5 meters:
   width of Lake Erie

10.2 Product of Powers Property
For use with Activity 10.2

Essential Question How can you use inductive reasoning to observe patterns and write general rules involving properties of exponents?

1 ACTIVITY: Finding Products of Powers

Work with a partner.

a. Complete the table.

Product	Repeated Multiplication Form	Power
$2^2 \cdot 2^4$	2222 4 · 16 6 4	64
$(-3)^2 \cdot (-3)^4$	5a	33
$7^3 \cdot 7^2$	78	91
$5.1^1 \cdot 5.1^6$	67 33	33
$(-4)^2 \cdot (-4)^2$	889 4,	33
$10^3 \cdot 10^5$	756	22
$\left(\dfrac{1}{2}\right)^5 \cdot \left(\dfrac{1}{2}\right)^5$	0046	33

b. **INDUCTIVE REASONING** Describe the pattern in the table. Then write a *general rule* for multiplying two powers that have the same base.

$$a^m \cdot a^n = a\text{——}$$

c. Use your rule to simplify the products in the first column of the table above. Does your rule give the results in the third column?

d. Most calculators have *exponent* keys that are used to evaluate powers. Use a calculator with an exponent key to evaluate the products in part (a).

10.2 Product of Powers Property (continued)

2 ACTIVITY: Writing a Rule for Powers of Powers

Work with a partner. Write the expression as a single power. Then write a *general rule* for finding a power of a power.

a. $\left(3^2\right)^3 =$

b. $\left(2^2\right)^4 =$

c. $\left(7^3\right)^2 =$

d. $\left(y^3\right)^3 =$

e. $\left(x^4\right)^2 =$

3 ACTIVITY: Writing a Rule for Powers of Products

Work with a partner. Write the expression as the product of two powers. Then write a *general rule* for finding a power of a product.

a. $\left(2 \bullet 3\right)^3 =$

b. $\left(2 \bullet 5\right)^2 =$

c. $\left(5 \bullet 4\right)^3 =$

d. $\left(6a\right)^4 =$

e. $\left(3x\right)^2 =$

10.2 **Product of Powers Property** (continued)

4 **ACTIVITY:** The Penny Puzzle

Work with a partner.

- The rows y and columns x of a chess board are numbered as shown.

- Each position on the chess board has a stack of pennies. (Only the first row is shown.)

- The number of pennies in each stack is $2^x \cdot 2^y$.

a. How many pennies are in the stack in location (3, 5)?

b. Which locations have 32 pennies in their stacks?

c. How much money (in dollars) is in the location with the tallest stack?

d. A penny is about 0.06 inch thick. About how tall (in inches) is the tallest stack?

What Is Your Answer?

5. **IN YOUR OWN WORDS** How can you use inductive reasoning to observe patterns and write general rules involving properties of exponents?

10.2 Practice
For use after Lesson 10.2

Simplify the expression. Write your answer as a power.

1. $(-6)^5 \cdot (-6)^4$

2. $x^1 \cdot x^9$

3. $\left(\dfrac{4}{5}\right)^3 \cdot \left(\dfrac{4}{5}\right)^{12}$

4. $(-1.5)^{11} \cdot (-1.5)^{11}$

5. $\left(y^{10}\right)^{20}$

6. $\left(\left(-\dfrac{2}{9}\right)^8\right)^7$

Simplify the expression.

7. $(2a)^6$

8. $(-4b)^4$

9. $\left(-\dfrac{9}{10}p\right)^2$

10. $(xy)^{15}$

11. $10^5 \cdot 10^3 - \left(10^1\right)^8$

12. $7^2\left(7^4 \cdot 7^4\right)$

13. The surface area of the Sun is about $4 \times 3.141 \times \left(7 \times 10^5\right)^2$ square kilometers. Simplify the expression.

Name_____ Date_____

10.3 Quotient of Powers Property
For use with Activity 10.3

Essential Question How can you divide two powers that have the same base?

1 ACTIVITY: Finding Quotients of Powers

Work with a partner.

a. Complete the table.

Quotient	Repeated Multiplication Form	Power
$\dfrac{2^4}{2^2}$		
$\dfrac{(-4)^5}{(-4)^2}$		
$\dfrac{7^7}{7^3}$		
$\dfrac{8.5^9}{8.5^6}$		
$\dfrac{10^8}{10^5}$		
$\dfrac{3^{12}}{3^4}$		
$\dfrac{(-5)^7}{(-5)^5}$		
$\dfrac{11^4}{11^1}$		

b. INDUCTIVE REASONING Describe the pattern in the table. Then write a rule for dividing two powers that have the same base.

$$\frac{a^m}{a^n} = a^{\underline{\quad}}$$

10.3 **Quotient of Powers Property** (continued)

c. Use your rule to simplify the quotients in the first column of the table on the previous page. Does your rule give the results in the third column?

2 **ACTIVITY:** Comparing Volumes

Work with a partner.

How many of the smaller cubes will fit inside the larger cube? Record your results in the table on the next page. Describe the pattern in the table.

a.

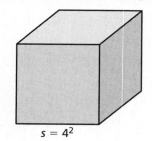

$s = 4$ $s = 4^2$

b.

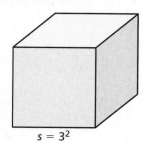

$s = 3$ $s = 3^2$

c.

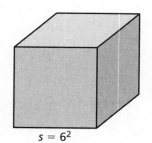

$s = 6$ $s = 6^2$

d.

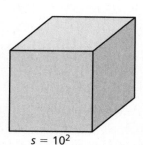

$s = 10$ $s = 10^2$

Name_____ Date_____

10.4 Zero and Negative Exponents
For use with Activity 10.4

Essential Question How can you evaluate a nonzero number with an exponent of zero? How can you evaluate a nonzero number with a negative integer exponent?

1 ACTIVITY: Using the Quotient of Powers Property

Work with a partner.

a. Complete the table.

Quotient	Quotient of Powers Property	Power
$\dfrac{5^3}{5^3}$		
$\dfrac{6^2}{6^2}$		
$\dfrac{(-3)^4}{(-3)^4}$		
$\dfrac{(-4)^5}{(-4)^5}$		

b. REPEATED REASONING Evaluate each expression in the first column of the table. What do you notice?

c. How can you use these results to define a^0 where $a \neq 0$?

10.4 Zero and Negative Exponents (continued)

2 **ACTIVITY:** Using the Product of Powers Property

Work with a partner.

a. Complete the table.

Product	Product of Powers Property	Power
$3^0 \bullet 3^4$		
$8^2 \bullet 8^0$		
$(-2)^3 \bullet (-2)^0$		
$\left(-\dfrac{1}{3}\right)^0 \bullet \left(-\dfrac{1}{3}\right)^5$		

b. Do these results support your definition in Activity 1(c)?

3 **ACTIVITY:** Using the Product of Powers Property

Work with a partner.

a. Complete the table.

Product	Product of Powers Property	Power
$5^{-3} \bullet 5^3$		
$6^2 \bullet 6^{-2}$		
$(-3)^4 \bullet (-3)^{-4}$		
$(-4)^{-5} \bullet (-4)^5$		

b. According to your results from Activities 1 and 2, the products in the first column are equal to what value?

10.4 **Zero and Negative Exponents** (continued)

  c. **REASONING** How does the Multiplicative Inverse Property help you to rewrite the numbers with negative exponents?

  d. **STRUCTURE** Use these results to define a^{-n} where $a \neq 0$ and n is an integer.

4 **ACTIVITY:** Using a Place Value Chart

Work with a partner. Use the place value chart that shows the number 3452.867.

Place Value Chart

thousands	hundreds	tens	ones	and	tenths	hundredths	thousandths
10^3	10^2	10^1	$10^\square$		$10^\square$	$10^\square$	$10^\square$
3	4	5	2	.	8	6	7

  a. **REPEATED REASONING** What pattern do you see in the exponents? Continue the pattern to find the other exponents.

  b. **STRUCTURE** Show how to write the expanded form of 3452.867.

What Is Your Answer?

  5. **IN YOUR OWN WORDS** How can you evaluate a nonzero number with an exponent of zero? How can you evaluate a nonzero number with a negative integer exponent?

10.4 Practice
For use after Lesson 10.4

Evaluate the expression.

1. 29^0

2. 12^{-1}

3. $10^{-4} \cdot 10^{-6}$

4. $\dfrac{1}{3^{-3}} \cdot \dfrac{1}{3^5}$

Simplify. Write the expression using only positive exponents.

5. $19x^{-6}$

6. $\dfrac{14a^{-5}}{a^{-8}}$

7. $3t^6 \cdot 8t^{-6}$

8. $\dfrac{12s^{-1} \cdot 4^{-2} \cdot r^3}{s^2 \cdot r^5}$

9. The density of a proton is about $\dfrac{1.64 \times 10^{-24}}{3.7 \times 10^{-38}}$ grams per cubic centimeter. Simplify the expression.

10.5 Reading Scientific Notation
For use with Activity 10.5

Essential Question How can you read numbers that are written in scientific notation?

1 ACTIVITY: Very Large Numbers

Work with a partner.

- Use a calculator. Experiment with multiplying large numbers until your calculator displays an answer that is *not* in standard form.

- When the calculator at the right was used to multiply 2 billion by 3 billion, it listed the result as

    $6.0E{+}18$.

- Multiply 2 billion by 3 billion by hand. Use the result to explain what $6.0E{+}18$ means.

- Check your explanation using products of other large numbers.

- Why didn't the calculator show the answer in standard form?

- Experiment to find the maximum number of digits your calculator displays. For instance, if you multiply 1000 by 1000 and your calculator shows 1,000,000, then it can display 7 digits.

10.5 **Reading Scientific Notation** (continued)

2 **ACTIVITY:** Very Small Numbers

Work with a partner.

- Use a calculator. Experiment with multiplying very small numbers until your calculator displays an answer that is *not* in standard form.

- When the calculator at the right was used to multiply 2 billionths by 3 billionths, it listed the result as

    $6.0E{-}18$.

- Multiply 2 billionths by 3 billionths by hand. Use the result to explain what $6.0E{-}18$ means.

- Check your explanation by calculating the products of other very small numbers.

3 **ACTIVITY:** Powers of 10 Matching Game

Work with a partner. Match each picture with its power of 10. Explain your reasoning.

| 10^5 m | 10^2 m | 10^0 m | 10^{-1} m | 10^{-2} m | 10^{-5} m |

A.

B.

C.

D.

E.

F.

10.6 Writing Scientific Notation
For use with Activity 10.6

Essential Question How can you write a number in scientific notation?

---

1 ACTIVITY: Finding pH Levels

Work with a partner. In chemistry, pH is a measure of the activity of dissolved hydrogen ions (H^+). Liquids with low pH values are called *acids*. **Liquids with high pH values are called** *bases*.

Find the pH of each liquid. Is the liquid a base, neutral, or an acid?

a. Lime juice: $[H^+] = 0.01$

b. Egg: $[H^+] = 0.00000001$

c. Distilled water: $[H^+] = 0.0000001$

d. Ammonia water:
$[H^+] = 0.00000000001$

e. Tomato juice: $[H^+] = 0.0001$

f. Hydrochloric acid: $[H^+] = 1$

pH	$[H^+]$
14	1×10^{-14}
13	1×10^{-13}
12	1×10^{-12}
11	1×10^{-11}
10	1×10^{-10}
9	1×10^{-9}
8	1×10^{-8}
7	1×10^{-7}
6	1×10^{-6}
5	1×10^{-5}
4	1×10^{-4}
3	1×10^{-3}
2	1×10^{-2}
1	1×10^{-1}
0	1×10^{0}

Bases

Neutral

Acids

10.6 Writing Scientific Notation (continued)

2 ACTIVITY: Writing Scientific Notation

Work with a partner. Match each planet with its distance from the Sun. Then write each distance in scientific notation. Do you think it is easier to match the distances when they are written in standard form or in scientific notation? Explain.

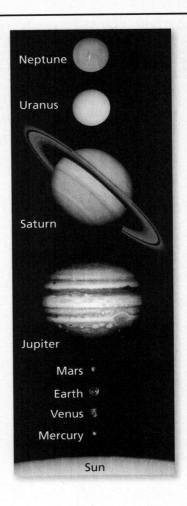

a. 1,800,000,000 miles

b. 67,000,000 miles

c. 890,000,000 miles

d. 93,000,000 miles

e. 140,000,000 miles

f. 2,800,000,000 miles

g. 480,000,000 miles

h. 36,000,000 miles

10.6 **Writing Scientific Notation** (continued)

3 ACTIVITY: Making a Scale Drawing

Work with a partner. The illustration in Activity 2 is not drawn to scale. Use the instructions below to make a scale drawing of the distances in our solar system.

- Cut a sheet of paper into three strips of equal width. Tape the strips together.

- Draw a long number line. Label the number line in hundreds of millions of miles.

- Locate each planet's position on the number line.

What Is Your Answer?

4. **IN YOUR OWN WORDS** How can you write a number in scientific notation?

Name _____ Date _____

Write the number in scientific notation.

1. 4,200,000

2. 0.038

3. 600,000

4. 0.0000808

5. 0.0007

6. 29,010,000,000

Order the numbers from least to greatest.

7. $6.4 \times 10^8, 5.3 \times 10^9, 2.3 \times 10^8$

8. $9.1 \times 10^{-3}, 9.6 \times 10^{-3}, 9.02 \times 10^{-3}$

9. $7.3 \times 10^7, 5.6 \times 10^{10}, 3.7 \times 10^9$

10. $1.4 \times 10^{-5}, 2.01 \times 10^{-15}, 6.3 \times 10^{-2}$

11. A patient has 0.0000075 gram of iron in 1 liter of blood. The normal level is between 6×10^{-7} gram and 1.6×10^{-5} gram. Is the patient's iron level normal? Write the patient's amount of iron in scientific notation.

10.7 Operations in Scientific Notation
For use with Activity 10.7

Essential Question How can you perform operations with numbers written in scientific notation?

---

1 **ACTIVITY:** Adding Numbers in Scientific Notation

Work with a partner. Consider the numbers 2.4×10^3 and 7.1×10^3.

a. Explain how to use order of operations to find the sum of these numbers. Then find the sum.

$$2.4 \times 10^3 + 7.1 \times 10^3$$

b. The factor _____ is common to both numbers. How can you use the Distributive Property to rewrite the sum $(2.4 \times 10^3) + (7.1 \times 10^3)$?

$(2.4 \times 10^3) + (7.1 \times 10^3) =$ _____   Distributive Property

c. Use order of operations to evaluate the expression you wrote in part (b). Compare the result with your answer in part (a).

d. **STRUCTURE** Write a rule you can use to add numbers written in scientific notation where the powers of 10 are the same. Then test your rule using the sums below.

• $(4.9 \times 10^5) + (1.8 \times 10^5) =$ _____

• $(3.85 \times 10^4) + (5.72 \times 10^4) =$ _____

---

2 **ACTIVITY:** Adding Numbers in Scientific Notation

Work with a partner. Consider the numbers 2.4×10^3 and 7.1×10^4.

a. Explain how to use order of operations to find the sum of these numbers. Then find the sum.

$$2.4 \times 10^3 + 7.1 \times 10^4$$

10.7 **Operations in Scientific Notation** (continued)

b. How is this pair of numbers different from the pair of numbers in Activity 1?

c. Explain why you cannot immediately use the rule you wrote in Activity 1(d) to find this sum.

d. STRUCTURE How can you rewrite one of the numbers so that you can use the rule you wrote in Activity 1(d)? Rewrite one of the numbers. Then find the sum using your rule and compare the result with your answer in part (a).

e. REASONING Does this procedure work when subtracting numbers written in scientific notation? Justify your answer by evaluating the differences below.

- $(8.2 \times 10^5) - (4.6 \times 10^5) = $ _____

- $(5.88 \times 10^5) - (1.5 \times 10^4) = $ _____

3 **ACTIVITY: Multiplying Numbers in Scientific Notation**

Work with a partner. Match each step with the correct description.

Step		Description
$(2.4 \times 10^3) \times (7.1 \times 10^3)$		**Original expression**
1.	$= 2.4 \times 7.1 \times 10^3 \times 10^3$	**A.** Write in standard form.
2.	$= (2.4 \times 7.1) \times (10^3 \times 10^3)$	**B.** Product of Powers Property
3.	$= 17.04 \times 10^6$	**C.** Write in scientific notation.
4.	$= 1.704 \times 10^1 \times 10^6$	**D.** Commutative Property of Multiplication
5.	$= 1.704 \times 10^7$	**E.** Simplify.
6.	$= 17,040,000$	**F.** Associative Property of Multiplication

Glossary

This student friendly glossary is designed to be a reference for key vocabulary, properties, and mathematical terms. Several of the entries include a short example to aid your understanding of important concepts.

Also available at *BigIdeasMath.com*:

- multi-language glossary
- vocabulary flash cards

Addition Property of Equality

Adding the same number to each side of an equation produces an equivalent equation.

$$
\begin{aligned}
x - 7 &= -6 \\
\underline{+7 \quad} &\underline{\quad +7} \\
x &= \quad 1
\end{aligned}
$$

adjacent angles

Two angles that share a common side and have the same vertex

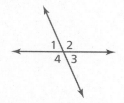

∠1 and ∠2 are adjacent.

∠2 and ∠4 are not adjacent.

---

angle of rotation

The number of degrees a figure rotates

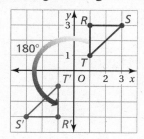

$\triangle RST$ has been rotated 180° to $\triangle R'S'T'$.

base (of a power)

The base of a power is the common factor.

See power.

---

center of dilation

A point with respect to which a figure is dilated

See dilation.

center of rotation

A point about which a figure is rotated

See rotation.

---

center of a sphere

The point inside a sphere that is the same distance from all points on the sphere

See sphere.

concave polygon

A polygon in which at least one line segment connecting any two vertices lies outside the polygon

congruent angles

Angles that have the same measure

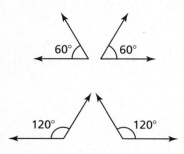

congruent figures

Figures that have the same size and the same shape

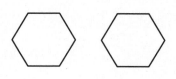

congruent sides

Sides that have the same length

Side *AB* and side *FG* are congruent sides.

convex polygon

A polygon in which every line segment connecting any two vertices lies entirely inside the polygon

coordinate plane

A coordinate plane is formed by the intersection of a horizontal number line and a vertical number line.

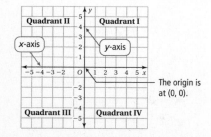

corresponding angles

Matching angles of two congruent figures

$\triangle ABC \cong \triangle DEF$

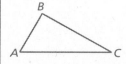 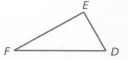

Corresponding angles: $\angle A$ and $\angle D$

$\angle B$ and $\angle E$

$\angle C$ and $\angle F$

corresponding sides

Matching sides of two congruent figures

$\triangle ABC \cong \triangle DEF$

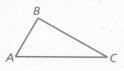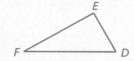

Corresponding sides: side AB and side DE
                  side BC and side EF
                  side AC and side DF

cube root

A number that, when multiplied by itself, and then multiplied by itself again, equals a given number

$$\sqrt[3]{8} = 2$$
$$\sqrt[3]{-27} = -3$$

degree

A unit used to measure angles

$$90°, 45°, 32°$$

dependent variable

The variable whose value depends on the independent variable in an equation in two variables

In the equation $y = 5x - 8$, y is the dependent variable.

dilation

A transformation in which a figure is made larger or smaller with respect to a fixed point called the center of dilation

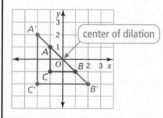

$A'B'C'$ is a dilation of ABC with respect to the origin. The scale factor is 2.

distance formula

The distance d between any two points (x_1, y_1) and (x_2, y_2) is given by the formula

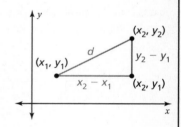

$$d = \sqrt{(x_2 - x_1)^2 + (y_2 - y_1)^2}.$$

Division Property of Equality

Dividing each side of an equation by the same number produces an equivalent equation.

$$4x = -40$$
$$\frac{4x}{4} = \frac{-40}{4}$$
$$x = -10$$

enlargement

A dilation with a scale factor greater than 1

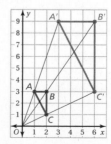

$A'B'C'$ is an enlargement of ABC.

equation	**equivalent equations**
A mathematical sentence that uses an equal sign to show that two expressions are equal $4x = 16, a + 7 = 21$	Equations that have the same solutions $2x - 8 = 0$ and $2x = 8$

evaluate (an algebraic expression)	**exponent**
Substitute a number for each variable in an algebraic expression. Then use the order of operations to find the value of the numerical expression. Evaluate $3x + 5$ when $x = 6$. $3x + 5 = 3(6) + 5$ $\quad\quad\quad = 18 + 5$ $\quad\quad\quad = 23$	The exponent of a power indicates the number of times a base is used as a factor. *See power.*

expression	**exterior angles**
A mathematical phrase containing numbers, operations, and/or variables $12 + 6, 18 + 3 \times 4,$ $8 + x, 6 \times a - b$	When two parallel lines are cut by a transversal, four exterior angles are formed on the outside of the parallel lines. 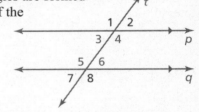 $\angle 3, \angle 4, \angle 5,$ and $\angle 6$ are interior angles. $\angle 1, \angle 2, \angle 7,$ and $\angle 8$ are exterior angles.

exterior angles of a polygon	**factor**
The angles outside a polygon that are adjacent to the interior angles 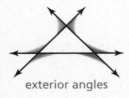 exterior angles	When whole numbers other than zero are multiplied together, each number is a factor of the product. $2 \times 3 \times 4 = 24$, so 2, 3, and 4 are factors of 24.

function	**function rule**
A relation that pairs each input with exactly one output The ordered pairs $(0, 1)$, $(1, 2)$, $(2, 4)$, and $(3, 6)$ represent a function. <table><tr><td>Ordered Pairs</td><td>Input</td><td>Output</td></tr><tr><td>(0, 1) (1, 2) (2, 4) (3, 6)</td><td>0 1 2 3</td><td>1 2 4 6</td></tr></table>	An equation that describes the relationship between inputs (independent variable) and outputs (dependent variable) The function rule "The output is three less than the input" is represented by the equation $y = x - 3$.
hemisphere	**hypotenuse**
One-half of a sphere 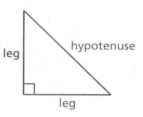	The side of a right triangle that is opposite the right angle leg hypotenuse leg
image	**independent variable**
The new figure formed by a transformation *See translation, reflection, rotation, and dilation.*	The variable representing the quantity that can change freely in an equation in two variables In the equation $y = 5x - 8$, x is the independent variable.
indirect measurement	**input**
Indirect measurement uses similar figures to find a missing measure when it is difficult to find directly. 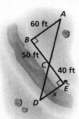 $$\frac{x}{60} = \frac{40}{50}$$ $$60 \cdot \frac{x}{60} = 60 \cdot \frac{40}{50}$$ $$x = 48$$ The distance across the river is 48 feet.	In a relation, inputs are associated with outputs. inputs 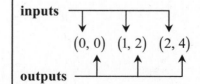(0, 0) (1, 2) (2, 4) outputs <table><tr><td>Input</td><td>Output</td></tr><tr><td>0 1 2</td><td>0 2 4</td></tr></table>

integers The set of whole numbers and their opposites $$\dots -3, -2, -1, 0, 1, 2, 3, \dots$$	**interior angles** When two parallel lines are cut by a transversal, four interior angles are formed on the inside of the parallel lines. *See exterior angles.*
interior angles of a polygon The angles inside a polygon 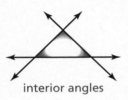 interior angles	**irrational number** A number that cannot be written as the ratio of two integers $$\pi, \sqrt{14}$$
joint frequency Each entry in a two-way table 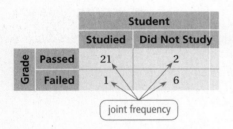	**legs** The two sides of a right triangle that form the right angle *See hypotenuse.*
line of best fit A precise line of fit that best models a set of data 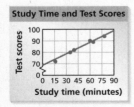	**line of fit** A line drawn on a scatter plot close to most of the data points; It can be used to estimate data on a graph. 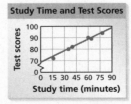

line of reflection

A line that a figure is reflected in to create a mirror image of the original figure

See reflection.

linear equation

An equation whose graph is a line

$y = x - 1$

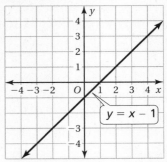

linear function

A function whose graph is a nonvertical line; a function that has a constant rate of change

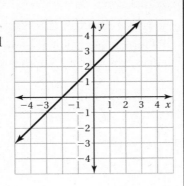

literal equation

An equation that has two or more variables

$$2y + 6x = 12$$

mapping diagram

A way to represent a relation

Input	Output
1	3
2	4
3	5
4	6

marginal frequencies

The sums of the rows and columns in a two-way table

		Age		
		12–13	**14–15**	**16–17**	**Total**
Student	**Rides Bus**	24	12	14	50
	Does Not Ride Bus	16	13	21	50
	Total	40	25	35	100

Multiplication Property of Equality

Multiplying each side of an equation by the same number produces an equivalent equation.

$$-\frac{2}{3}x = 8$$
$$-\frac{3}{2} \cdot \left(-\frac{2}{3}x\right) = -\frac{3}{2} \cdot 8$$
$$x = -12$$

nonlinear function

A function that does not have a constant rate of change; a function whose graph is not a line

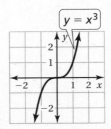

$y = x^3$

ordered pair	**origin**
A pair of numbers (x, y) used to locate a point in a coordinate plane; The first number is the x-coordinate, and the second number is the y-coordinate. The x-coordinate of the point $(-2, 1)$ is -2, and the y-coordinate is 1.	The point, represented by the ordered pair $(0, 0)$, where the horizontal and the vertical number lines intersect in a coordinate plane *See coordinate plane.*
output	**parallel lines**
In a relation, inputs are associated with outputs. *See input.*	Lines in the same plane that do not intersect; Nonvertical parallel lines have the same slope. All vertical lines are parallel. 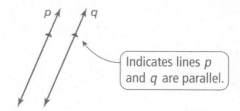 Indicates lines p and q are parallel.
perfect cube	**perfect square**
A number that can be written as the cube of an integer $$-27, 8, 125$$	A number with integers as its square roots $$16, 25, 81$$
perpendicular lines	**point-slope form**
Lines in the same plane that intersect at right angles; Two nonvertical lines are perpendicular when the product of their slopes is -1. Vertical lines are perpendicular to horizontal lines. 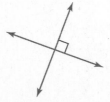	A linear equation written in the form $y - y_1 = m(x - x_1)$ is in point-slope form. The line passes through the point (x_1, y_1), and the slope of the line is m. $$y - 1 = \frac{2}{3}(x + 6)$$

polygon

A closed figure in a plane that is made up of three or more line segments that intersect only at their endpoints

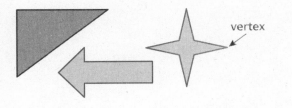

vertex

power

A product of repeated factors

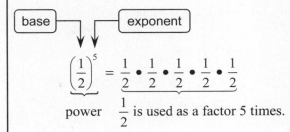

base   exponent

$$\underbrace{\left(\frac{1}{2}\right)^5}_{\text{power}} = \underbrace{\frac{1}{2} \cdot \frac{1}{2} \cdot \frac{1}{2} \cdot \frac{1}{2} \cdot \frac{1}{2}}$$

power $\frac{1}{2}$ is used as a factor 5 times.

Power of a Power Property

To find a power of a power, multiply the exponents.

$$\left(3^4\right)^2 = 3^{4 \cdot 2} = 3^8$$

$$\left(a^m\right)^n = a^{mn}$$

Power of a Product Property

To find a power of a product, find the power of each factor and multiply.

$$\left(5 \cdot 7\right)^4 = 5^4 \cdot 7^4$$

$$\left(ab\right)^m = a^m b^m$$

Product of Powers Property

To multiply powers with the same base, add their exponents.

$$3^7 \cdot 3^{10} = 3^{7+10} = 3^{17}$$

$$a^m \cdot a^n = a^{m+n}$$

proportion

An equation stating that two ratios are equivalent

$$\frac{3}{4} = \frac{12}{16}$$

proportional

Two quantities that form a proportion are proportional.

Because $\frac{3}{4}$ and $\frac{12}{16}$ form a proportion,

$\frac{3}{4}$ and $\frac{12}{16}$ are proportional.

Pythagorean Theorem

In any right triangle, the sum of the squares of the lengths of the legs is equal to the square of the length of the hypotenuse.

$$a^2 + b^2 = c^2$$

13 cm   5 cm   12 cm

$$5^2 + 12^2 = 13^2$$

Quotient of Powers Property

To divide powers with the same base, subtract their exponents.

$$\frac{9^7}{9^3} = 9^{7-3} = 9^4$$

$$\frac{a^m}{a^n} = a^{m-n}, \text{ where } a \neq 0$$

radical sign

The symbol $\sqrt{}$ which is used to represent a square root

$$\sqrt{25} = 5$$
$$-\sqrt{49} = -7$$
$$\pm\sqrt{100} = \pm 10$$

radicand

The number under a radical sign

The radicand of $\sqrt{25}$ is 25.

radius of a sphere

The distance from the center of a sphere to any point on the sphere

See sphere.

rate

A ratio of two quantities with different units

You read 3 books every 2 weeks.

ratio

A comparison of two quantities using division; The ratio of a to b (where $b \neq 0$) can be written as a to b, $a : b$, or $\frac{a}{b}$.

$$4 \text{ to } 1, 4 : 1, \text{ or } \frac{4}{1}$$

rational number

A number that can be written as $\frac{a}{b}$ where a and b are integers and $b \neq 0$

$$3 = \frac{3}{1}, \qquad\qquad -\frac{2}{5} = \frac{-2}{5}$$

$$0.25 = \frac{1}{4}, \qquad\qquad 1\frac{1}{3} = \frac{4}{3}$$

real numbers

The set of all rational and irrational numbers

$$4, -6.5, \pi, \sqrt{14}$$

reduction

A dilation with a scale factor greater than 0 and less than 1

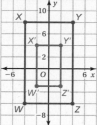

W′X′Y′Z′ is a reduction of *WXYZ*.

reflection

A transformation in which a figure is reflected in a line called the line of reflection; A reflection creates a mirror image of the original figure.

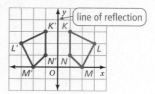

K′L′M′N′ is a reflection of *KLMN* over the *y*-axis.

regular polygon

A polygon in which all the sides are congruent, and all the interior angles are congruent

relation

A relation pairs inputs with outputs and can be represented by ordering pairs on a mapping diagram.

Ordered Pairs
(0, 1)
(1, 2)
(2, 4)

Mapping Diagram

right angle

An angle whose measure is 90°

right triangle

A triangle that has one right angle

rise

The change in *y* between any two points on a line

See slope.

rotation

A transformation in which a figure is rotated about a point called the center of rotation; The number of degrees a figure rotates is the angle of rotation.

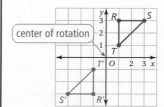

△*RST* has been rotated about the origin *O* to △*R′S′T′*.

run The change in x between any two points on a line *See slope.*	**scale factor (of a dilation)** The ratio of the side lengths of the image of a dilation to the corresponding side lengths of the original figure *See dilation.*
scatter plot A graph that shows the relationship between two data sets using ordered pairs in a coordinate plane 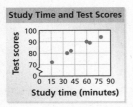	**scientific notation** A number is written in scientific notation when it is represented as the product of a factor and a power of 10. The factor must be greater than or equal to 1 and less than 10. $$8.3 \times 10^4$$ $$4 \times 10^{-3}$$
similar figures Figures that have the same shape but not necessarily the same size; Two figures are similar when corresponding side lengths are proportional and corresponding angles are congruent. 	**similar solids** Solids that have the same shape and proportional corresponding dimensions 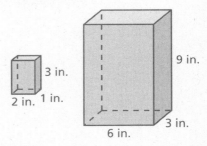
slope The slope m of a line is a ratio of the change in y (the rise) to the change in x (the run) between any two points (x_1, y_1) and (x_2, y_2) on a line. It is a measure of the steepness of a line. $$m = \frac{\text{rise}}{\text{run}} = \frac{\text{change in } y}{\text{change in } x}$$ $$= \frac{y_2 - y_1}{x_2 - x_1}$$ 	**slope-intercept form** A linear equation written in the form $y = mx + b$ is in slope-intercept form. The slope of the line is m, and the y-intercept of the line is b. The slope is 1 and the y-intercept is 2. 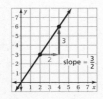

solid A three-dimensional figure that encloses a space 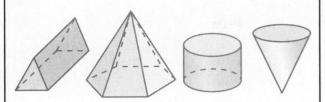	**solution of an equation** A value that makes an equation true 6 is the solution of the equation $x - 4 = 2$.
solution of a linear equation All of the points on a line	**solution of a system of linear equations** An ordered pair that is a solution of each equation in a system $(1, -3)$ is the solution of the following system of linear equations. $$4x - y = 7$$ $$2x + 3y = -7$$
sphere The set of all points in space that are the same distance from a point called the center 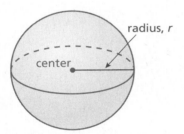	**square root** A number that, when multiplied by itself, equals a given number The two square roots of 100 are 10 and -10. $\pm\sqrt{100} = \pm 10$
standard form The standard form of a linear equation is $ax + by = c$, where a and b are not both zero. $$-2x + 3y = -6$$	**Subtraction Property of Equality** Subtracting the same number from each side of an equation produces an equivalent equation. $$x + 10 = -12$$ $$\underline{-\ 10 \qquad -\ 10}$$ $$x = -22$$

system of linear equations	**theorem**
A set of two or more linear equations in the same variables, also called a linear system. $\begin{array}{ll} y = x + 1 & \text{Equation 1} \\ y = 2x - 7 & \text{Equation 2} \end{array}$	A rule in mathematics The Pythagorean Theorem

transformation	**translation**
A transformation changes a figure into another figure. *See translation, reflection, rotation, and dilation.*	A transformation in which a figure slides but does not turn; Every point of the figure moves the same distance and in the same direction. 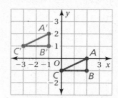 *ABC* has been translated 3 units left and 2 units up to $A'B'C'$.

transversal	**two-way table**
A line that intersects two or more lines 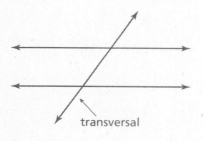	Displays two categories of data collected from the same source 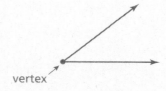

variable	**vertex (of an angle)**
A symbol that represents one or more numbers x is a variable in $2x + 1$.	The point at which the two sides of an angle meet

vertex (of a polygon) A point at which two sides of a polygon meet; The plural of vertex is vertices. *See polygon.*	**whole numbers** The numbers 0, 1, 2, 3, 4, …
x-axis The horizontal number line in a coordinate plane *See coordinate plane.*	**x-coordinate** The first coordinate in an ordered pair, which indicates how many units to move to the left or right from the origin In the ordered pair $(3, 5)$, the x-coordinate is 3.
x-intercept The x-coordinate of the point where a line crosses the x-axis 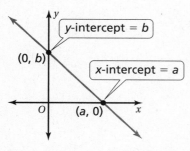	**y-axis** The vertical number line in a coordinate plane *See coordinate plane.*
y-coordinate The second coordinate in an ordered pair, which indicates how many units to move up or down from the origin In the ordered pair $(3, 5)$, the y-coordinate is 5.	**y-intercept** The y-coordinate of the point where a line crosses the y-axis *See x-intercept.*

Photo Credits

37 *top* ©iStockphoto.com/Viatcheslav Dusaleev; *bottom left* ©iStockphoto.com/Jason Mooy; *bottom right* ©iStockphoto.com/Felix Möckel; **49** Elena Elisseeva/Shutterstock.com.; **52** Estate Craft Homes, Inc.; **88** ©iStockphoto.com/biffspandex; **106** Talvi/Shutterstock.com; **134** ©iStockphoto.com/PeskyMonkey; **151** ©Oxford Science Archive/Heritage Images/Imagestate; **185** *baseball* Kittisak/Shutterstock.com; *golf ball* tezzstock/Shutterstock.com; *basketball* vasosh/Shutterstock.com; *tennis ball* UKRID/Shutterstock.com; *water polo ball* John Kasawa/Shutterstock.com; *softball* Ra Studio/Shutterstock.com; *volleyball* vberla/Shutterstock.com; **189** Gina Brockett; **198** Larry Korhnak; **199** Photo by Andy Newman; **204** ©iStockphoto.com/Franck Boston; **205** Stevyn Colgan; **219** ©iStockphoto.com/Kais Tolmats; **220** *top right* ©iStockphoto.com/Kais Tolmats; *Activity 3a and d* Tom C Amon/Shutterstock.com; *Activity 3b* Olga Gabay/Shutterstock.com; *Activity 3c* NASA/MODIS Rapid Response/Jeff Schmaltz; *Activity 3f* HuHu/Shutterstock.com; **221** *Activity 4a* PILart/Shutterstock.com; *Activity 4b* Matthew Cole/Shutterstock.com; *Activity 4c* Yanas/Shutterstock.com; *Activity 4e* unkreativ/Shutterstock.com; **224** NASA

Cartoon Illustrations: Tyler Stout

Cover Image: Pavelk/Shutterstock.com, Pincasso/Shutterstock.com, valdis torms/Shutterstock.com

a.

d.

Pattern Blocks – Chapter 2 Section 6*

*Available at *BigIdeasMath.com.*

*Available at *BigIdeasMath.com.*

Algebra Tiles*

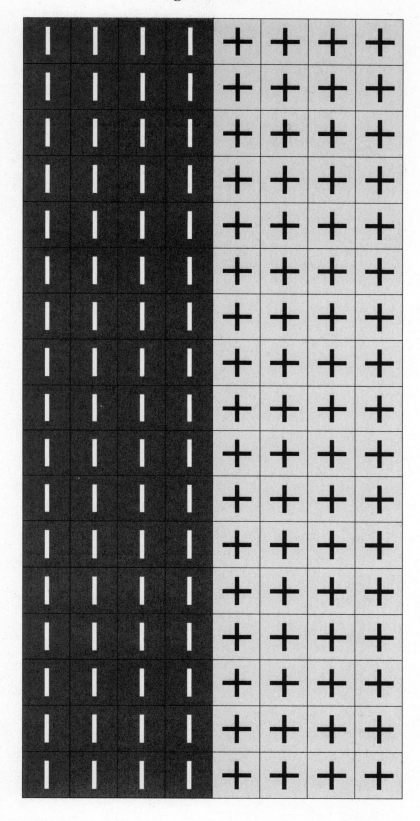

Algebra Tiles*

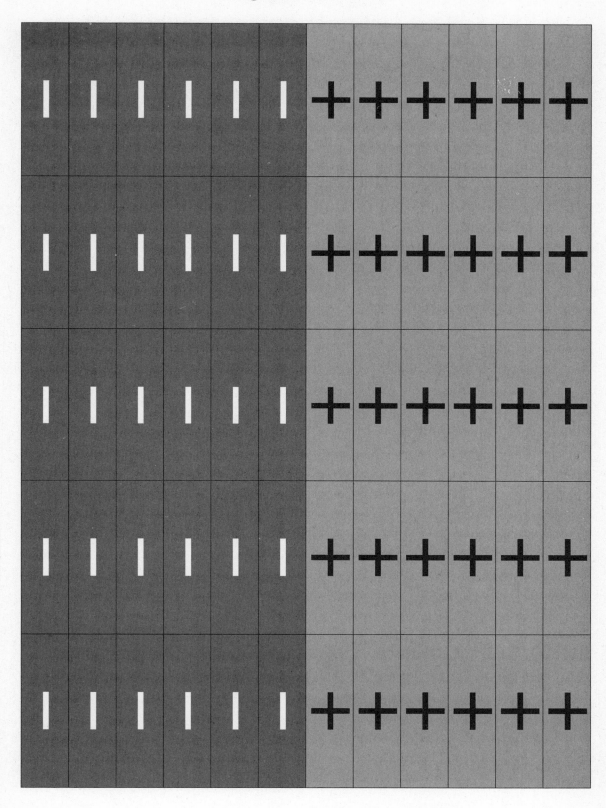

*Available at *BigIdeasMath.com*.